SELECTED APPLICATIONS
OF NONLINEAR PROGRAMMING

SELECTED APPLICATIONS OF NONLINEAR PROGRAMMING

Jerome Bracken
and
Garth P. McCormick

Research Analysis Corporation
McLean, Virginia

John Wiley & Sons, Inc., New York · London · Sydney · Toronto

Copyright © 1968 by Research Analysis Corporation. Reproduction in whole or in part is permitted for any purpose of the United States Government without payment of royalties.

Preparation of this book was undertaken in the course of research sponsored by contract DA-44-188-ARO-1 between the U.S. Army Research Office and Research Analysis Corporation.

Library of Congress Catalog Card Number: 68-26848
SBN 471 09440 4
Printed in the United States of America

To Marilyn and Virginia

PREFACE

This book presents selected applications of nonlinear programming in some detail. The first chapter, which is a general introduction to nonlinear programming, contains definitions, classification of problems, mathematical characteristics, and solution procedures. The remaining chapters deal with various problems and their nonlinear programming models. For most of the problems considered we give a brief summary, a mathematical formulation of a nonlinear programming model, and one or two examples. Some of the examples are based on "live" or real-world data, others on hypothetical data.

The *mathematical programming* problem is to determine values for a specified set of variables that *optimize* (maximize or minimize) a numerical function of the variables, subject to various constraint relations that are numerical functions of the variables.

When both the constraints and the objective function are linear, the problem is a *linear programming* problem. We have not dealt with linear programming models in this book because the area admits of a large quantity of literature and also because computer programs have been developed to such an extent that it is difficult to do justice to the topic of applying them to various problems without writing a great deal on each. A good introductory description of the application of the LP/90 linear programming simplex method computer program to optimizing production mix in an oil refinery is contained in Chapter 3 of *An Introduction to Linear Programming*, IBM Data Processing Application, International Business Machines Corporation, E20-8171, 1964.

When one or more of the constraints, or the objective function, are nonlinear, the problem is a *nonlinear programming* problem. Nonlinear programming problems may have widely varying characteristics, and certain limitations on the forms of the functions are necessary if problems are to be solved. We discuss mathematical limitations in Chapter 1. Research is progressing on many of the unsolved problems and therefore it is difficult to characterize the kinds of problem that can or cannot be solved. It is not our purpose to delimit the areas of possible nonlinear programming applications by completely characterizing the nonlinear programming problems that can be solved mathematically. Rather, we use the approach of discussing solved problems by assuming that model building will interact with mathematics to influence the direction of mathematical developments.

It should be noted that procedures used to solve nonlinear programming problems can in many cases solve linear programming problems. They are, however, seldom as efficient and should be used only when justifiable in the particular situation.

This book is intended primarily to help the interested reader to acquire facility in building mathematical programming models—in particular, nonlinear programming models. Mathematical programming as characterized above is not the context in which many are introduced to the field and thus is not the usual foundation for thinking about problem areas to which mathematical programming models might be profitably applied. Rather, many think of linear input-output systems with linear objective functions and try to extend the framework to more general problems. It is particularly difficult for persons with limited mathematical training, as is often the case in linear programming courses in business schools, to go beyond linear systems. We hope that this book will serve as a vehicle for extending the mathematical programming horizons of persons with such backgrounds and interests. Notes for the book have been used as supplementary material in a graduate course in mathematical programming, which emphasizes model building, in the School of Government and Business Administration of The George Washington University, with one of the authors as the teacher.

This book is also intended to be of interest to practicing model builders in that it presents, in one source at a consistent level, nonlinear programming models from diverse fields. The material is given in sufficient detail that there should be few questions about model formulation. Computational questions are not discussed in depth, but all of the application examples have been solved by the sequential unconstrained minimization technique for nonlinear programming (SUMT). A brief description of SUMT is given in Chapter 1 and further information about it can be found in the references cited in Chapter 1.

It should be noted that this book has been written in parallel with Anthony V. Fiacco and Garth P. McCormick: Nonlinear Programming: *Sequential Unconstrained Minimization Techniques.*

Preface

We should like to acknowledge Nicholas M. Smith for his continued personal interest and his effective technical direction of this work. He has been the supervisor of Research Analysis Corporation's nonlinear programming research program since its inception.

Sustained support of the work at RAC has come from the Army Research Office. Ralph T. Tierno, Colonel, USA, has been particularly helpful in expediting the publication of this book.

We also acknowledge our colleagues who worked with us on the models presented in this book: Richard M. Soland, Benjamin C. Rush, W. Charles Mylander, III, Dale M. Heien, Robert E. Pugh, Anthony V. Fiacco, and Arnold P. Jones. Their collaborations are noted in the individual chapters.

Special thanks for patient and accurate typing are due to Mrs. Ann E. Snyder.

JEROME BRACKEN
GARTH P. MCCORMICK

McLean, Virginia
May 1968

CONTENTS

1	**INTRODUCTION**	**1**
	1.1 Mathematical Programming—General	1
	1.2 Classification of Optimization Models	4
	1.3 Algebraic Structure of Mathematical Programming Problems	5
	1.4 The Sequential Unconstrained Minimization Technique for Nonlinear Programming	16
	1.5 References	19
2	**WEAPONS ASSIGNMENT**	**22**
	2.1 Model for Maximizing Expected Target Damage Value	22
	2.2 Model for Inflicting Specified Damage with Minimum Cost	23
	2.3 Example	24
	2.4 Source of Problem and Reference	27
3	**BID EVALUATION**	**28**
	3.1 Description of Bid Evaluation Problem	28
	3.2 Branch and Bound Method for Solution	30
	3.3 Solution of Example Problem	33
	3.4 Source of Problem and References	36
4	**ALKYLATION PROCESS OPTIMIZATION**	**37**
	4.1 Description of Alkylation Process Model	37
	4.2 Example of Alkylation Process Model Application	42
	4.3 References	45
5	**CHEMICAL EQUILIBRIUM**	**46**
	5.1 Chemical Equilibrium Model	46
	5.2 Example of Chemical Equilibrium Model Application	47
	5.3 Source of Problem and References	49

6	STRUCTURAL OPTIMIZATION	50
	6.1 Description of Design Problem	50
	6.2 Description of Objective Function	52
	6.3 Description of Constraints	53
	6.4 Specification of Nonlinear Programming Model	55
	6.5 Results of Solution	57
	6.6 References	57
7	LAUNCH VEHICLE DESIGN AND COSTING	58
	7.1 Introduction	58
	7.2 Development of Cost Function	60
	7.3 Development of Constraints	68
	7.4 Application of Model	77
	7.5 Source of Problem and References	81
8	PARAMETER ESTIMATION IN CURVE FITTING	83
	8.1 Linear Regression with Various Criteria	83
	8.2 Example of a Linear Regression Problem with Various Criteria	86
	8.3 Nonlinear Regression with Various Criteria	88
	8.4 Example of a Nonlinear Regression Problem	89
	8.5 Maximum Likelihood Estimation	90
	8.6 Source of Problem and References	93
9	DETERMINISTIC NONLINEAR PROGRAMMING EQUIVALENTS FOR STOCHASTIC LINEAR PROGRAMMING PROBLEMS	94
	9.1 Chance-Constrained Programming Problem and Its Deterministic Equivalent	94
	9.2 Example of Deterministic Equivalent for a Chance-Constrained Programming Problem	98
	9.3 Source of Problem and References	100
10	OPTIMAL SAMPLE SIZES IN STRATIFIED SAMPLING ON SEVERAL VARIATES	101
	10.1 Stratified Sampling Problems with Nonlinear Programming Models	102
	10.2 Example with Four Strata and Two Variates	103
	10.3 Source of Problem and References	105

INDEX   107

SELECTED APPLICATIONS OF NONLINEAR PROGRAMMING

I
INTRODUCTION

1.1 MATHEMATICAL PROGRAMMING—GENERAL

The problem of optimizing (either maximizing or minimizing) a numerical function of one or more variables subject to constraints on the variables is called the mathematical programming, or constrained optimization, problem. The general mathematical programming problem can be formulated as follows. Determine values for n variables $x = (x_1, \ldots, x_n)$ that optimize the function (called the objective or criterion function)

$$f(x) = f(x_1, \ldots, x_n) \tag{1.1}$$

subject to l linear and/or nonlinear inequality constraints

$$g_i(x) = g_i(x_1, \ldots, x_n) \geq 0, \qquad i = 1, \ldots, l \tag{1.2}$$

and to $m - l$ linear and/or nonlinear equality constraints

$$g_i(x) = g_i(x_1, \ldots, x_n) = 0, \qquad i = l+1, \ldots, m. \tag{1.3}$$

When the objective function (1.1) and the m constraints (1.2 and 1.3) are linear, the mathematical programming problem is a linear programming problem. When either the objective function or one or more of the constraints are nonlinear, the programming problem is a nonlinear programming problem.

In considering the nonlinear programming problem we prefer always to think of a minimizing problem, but this implies no loss of generality since, if the objective is to maximize $F(x)$, this can be converted to an equivalent minimization problem by letting $f(x) = -F(x)$ and minimizing $f(x)$. Also, the equal-to-or-greater-than inequality constraints $g_i(x) \geq 0$, $i = 1, \ldots, l$ do not impose any loss of generality since, if there is a constraint $G_i(x) \leq 0$, we may define $g_i(x) = -G_i(x)$ and write $g_i(x) \geq 0$.

The values of m and n need not be related. Some or all of the variables x_j ($j = 1, \ldots, n$) may be restricted as to sign, or have lower and/or upper bounds. Such constraints may be included in the g_i's. It is possible for there to be no constraints ($m = 0$), in which case the problem is called an unconstrained optimization problem. However, it may still be considered to be a mathematical programming problem.

Historically, we may characterize two groups of researchers interested in solving mathematical programming problems.

The first and older group is composed of physicists, chemists, scientists, and engineers. The problems of interest to this group generally contain highly nonlinear relationships. The engineering problems are usually ones of design: for example, the design of oil refineries to maximize return on capital subject to production requirements and technological limitations [25], and the design of structures to minimize total structure weight subject to load conditions, stress, displacement, and buckling limitations [40]. Problems in optimal control and the calculus of variations originate in this group. These problems usually involve minimizing an integral over time and hence do not fit into the format of the general mathematical programming problem stated above, but with appropriate analysis many of them can be represented as mathematical programming problems. For example, the problem of minimizing the total fuel expended by a space vehicle while maneuvering to satisfy certain trajectory requirements and under known equations of motion can be cast as a nonlinear programming problem [32].

Another important class of problems historically of interest in the physical sciences is that of nonlinear curve fitting. Longstanding attempts have been made to find parameter values that, with appropriate models, will explain scientifically observed data. Also, recent investigators in the fields of economics, psychology, and medicine have been using nonlinear statistical models, which may often be regarded as nonlinear curve-fitting models, to try to understand social phenomena. Examples are economic growth models [42] and investigation of important causes of coronary disease [44].

Several "classic" techniques for solving optimization problems have been motivated by physical science problems. The method of steepest descent developed by Cauchy in 1847 [6] for solving unconstrained optimization problems, and the method of Lagrange multipliers [13] for solving equality constrained optimization problems, are two of the best known techniques. The latter work foreshadowed much of modern duality theory [13, p. 232; 45]. Another technique that comes from developments in classical physics is the idea of a forcing function, or potential function, to solve equality-constrained problems [39].

The second group of people motivating work in general mathematical programming is composed of economists, operations research analysts, management scientists, and planners, who since the late 1930s have been involved in quantitative studies of how best to allocate scarce resources.

Since models of resource allocation can many times be reasonably formulated using only linear relationships, this group has been concerned mainly with linear programming models.

The simplex method, proposed by G. B. Dantzig in 1947 [14], is an extremely effective method for solving linear programming problems. Most of the effort in the past 20 years directed toward solving mathematical programming problems has been in the area of linear programming. Literally hundreds of articles and books have been written about linear programming problems and methods of solution. Computer programs that handle large linear problems efficiently have been coded. Recently much theoretical and computational work has been done on large linear problems that can be decomposed or partitioned into smaller linear optimization problems.

Linear programming has been successfully applied to problems of resource allocation in such industries as coal, iron and steel, paper, and petroleum; in transportation problems and network theory; in contract awards, production scheduling, structural design, and economic analysis; and in agricultural and military applications. The literature dealing with applications is extensive.

The success of the linear programming approach to modeling real-world problems, and the facility of solution of the models, has caused this second group of researchers to become interested in the general mathematical programming problem, in which the resource and objective functions may be nonlinearly related to activity levels. Concepts such as diminishing marginal costs lead to nonlinear cost functions, and models of supply and demand yield nonlinear profit functions. Many models of resource allocation problems have nonlinear aspects in their constraints and/or objective functions.

This book includes examples from both of these two traditions or groups. The chapters on alkylation, structural optimization, and curve fitting are primarily related to the physical sciences, whereas the weapons assignment, bid evaluation, cattle feed, and stratified sampling chapters deal with nonlinear resource allocation models. The chemical equilibrium and launch vehicle design and costing problems have elements of both traditions.

Finally, some mention sould be made of the relationships between mathematical programming and other applied mathematical tools. Early results established the connection between linear programming and two-person zero sum games [23]. Recent work has established the equivalence of certain mathematical programming problems with the findings of equilibrium points for nonzero sum two-person games [28, 29]. An interesting relationship between concave programming and n-person games is discussed in [37]. Applications to the solution of nonlinear equations are well known as a special case of nonlinear curve fitting.

1.2 CLASSIFICATION OF OPTIMIZATION MODELS

Any real-world optimization problem may be characterized by five qualities. The problem functions may all be *linear*, or some may be *nonlinear*. The functional relationships may be known (*deterministic*), or there may be uncertainty about them (*probabilistic*). The optimization may take place at a fixed point in time (*static*), or it may be an optimization over time (*dynamic*). The variables of the problem may be allowed to take on a spectrum of real values (*continuous*), or some or all may be required to be integers (*discrete*). Finally, the problem functions may all be continuously differentiable (*smooth*), or some may have points where the functions are nondifferentiable (*nonsmooth*).

The statement of the general mathematical programming problem allows for nonlinearities in the problem functions, but it assumes a deterministic model. That is, given x, the values assumed by the functions $f(x)$, $g_i(x)$, $i = 1, \ldots, m$ are uniquely defined. Also, the vector x is considered only at one point in time—the static situation. There is no way to restrict the solution vector to take on integer values for a selected subset of variables. Although there are algorithms for linear programming that generate integer solutions, at present none exists for nonlinear programming. In principle the cutting plane method referred to in the subsection on convex programming in Section 1.3, or Bellman's principle of optimality [2, 3], can be applied to integer programming problems. Finally, almost all algorithms for solving nonlinear programming problems require that the functions be smooth. Thus in most cases the model builder must develop deterministic-static-continuous-smooth models to represent probabilistic-dynamic-discrete-nonsmooth real-world problems.

In this book the weapons assignment, bid evaluation, and stratified sampling problems require integer solutions. Because of the large numbers involved (and in view of the accuracy of the inputs), rounding the fractional solution to the nearest integer provides a satisfactory resolution for these examples. This stratagem is not a good one for combinatorial problems (requiring 0-or-1 solutions) or for logical problems that have been converted to nonlinear integer programming problems. In these cases graph theory, dynamic programming, or combinatorial analysis is required.

All of the examples included in this book are static; they deal with the values of the variables at one point in time. In problems such as optimal control [38] or chance-constrained programming [10], in which a solution vector over time is required, the usual tactic is to divide the time interval into a fixed number of intervals and make each combination of variable-time points into a separate variable. This technique, of necessity, generates very large programming problems, and it is responsible for much recent linear programming work in large-scale decomposable systems. Very little work has been done in large-scale nonlinear programming on problems with special structures that result from methods of approximating optimization over time.

For problems with stochastic or probabilistic elements, a higher level of difficulty is introduced. Two questions must be resolved: (a) What constitutes a proper objective function for a problem whose real payoff has a probability distribution? (b) What does constraint satisfaction mean when constraint relationships are probabilistic? A common way of resolving the difficulty of specifying the proper objective function is seen in the weapons assignment problem. Here the outcome of the assignment of one variable to a target is known with a certain probability. The criterion function is given as the *expected* damage done to the target complex. In the cattle feed problem, where the nutrient content of each food is known to satisfy a certain probability distribution, the Charnes-Cooper concept [7, 9] of establishing a deterministic equivalent problem to the chance-constrained problem, whereby constraints are satisfied with a preassigned probability, is employed. In problems like the alkylation problem, where the functional relationships between independent and dependent variables as defined in the model are determined from experimental data and are represented in the model in algebraic form, one can use the results of nonlinear curve fitting as a deterministic input, or simply require (as is done here) that the relationships fall within bounds of specified experimental error.

For problems where the functions are not smooth, it is often possible to develop an equivalent sequence of models that have smooth functions. The bid evaluation problem is one such example, in which recent work in branch and bound techniques allows for efficient solutions.

Finally, although most of the problems in this book are not deterministic-static-continuous-smooth mathematical programming problems, the models that approximate them necessarily have these characteristics.

1.3 ALGEBRAIC STRUCTURE OF MATHEMATICAL PROGRAMMING PROBLEMS

Quite apart from the historical setting and disciplines from which the mathematical programming problems originated, and apart from their qualitative characteristics, there is a third way of describing them, which we shall call here their algebraic structure. This is the way in which mathematical programming problems are regarded by those who develop algorithms to solve them.

Some of the terms employed are linear, nonlinear, quadratic, convex, concave, and separable. We describe below the mathematical programming problems characterized by these terms and mention briefly the algorithms developed for their solution.

Linear Programming

Define

$$A = (a_{ij}), \quad i = 1, \ldots, m, \quad \text{and} \quad j = 1, \ldots, n,$$
$$b = (b_1, \ldots, b_m)^t,$$
$$c = (c_1, \ldots, c_n)^t.$$

The linear programming problem is to choose $x = (x_1, \ldots, x_n)^t$ to

$$\text{minimize } c^t x \tag{1.4}$$

subject to

$$Ax - b \geq 0, \quad x \geq 0. \tag{1.5}$$

The linear programming problem is the best known mathematical programming problem. It is characterized by a linear objective function, linear constraints, and non-negativity requirements on the variables. Any solution must (under mild nondegeneracy assumptions) lie on a vertex of the convex polyhedron described by constraints (1.5). Because of this, most methods for solving the linear programming problem are efficient computational schemes for moving from one vertex to an adjacent vertex in an effort to find the one that is optimal. Since there are a finite number of vertices, the algorithms using this property are guaranteed to yield the solution in a finite number of iterations. Once the appropriate vertex is found, the solution vector is uniquely determined by those equations that define that vertex. It is this property of the linear programming problem that makes it relatively easy to solve. For, from the vertex solution property, when m, the number of rows of the matrix A, is less than n, the number of variables (as is usually the case in practical problems), then at least $n - m$ of the variables of the solution vector are equal to zero. The simplex method for solving linear programming problems makes efficient use of this fact. The degree of difficulty in solving large linear programming problems is a function of m and of the density (per cent of nonzero elements) of the matrix A. Linear programming computer codes readily available for large-scale computers have been developed to solve problems with up to 4000 constraints and an unlimited number of variables.

It is not surprising that many early efforts to develop nonlinear programming algorithms were attempts to convert these problems to sequences of linear programming problems. Much of the effort today still proceeds from this point of view.

Structure of Mathematical Programming Problems

The following simple example gives some idea of the geometry of linear programming problems. The problem is to choose x_1 and x_2 to

$$\text{minimize } -3x_1 - 2x_2$$

subject to

$$-2x_1 - x_2 + 3 \geq 0,$$
$$-x_1 - x_2 + 2 \geq 0,$$
$$x_1 \geq 0, \quad x_2 \geq 0.$$

In Figure 1.1 the problem is depicted geometrically. The constraints of the problem define the feasible region, which is shaded. It is a bounded convex polyhedron in two dimensions. The dashed lines are isocontours of the objective function; that is, lines where the function $-3x_1 - 2x_2$ has constant value. The solution as obtained by inspection lies on the vertex $(1, 1)$. Geometrically any mathematical programming problem, and in particular this example linear programming problem, is to find the values of the variables associated with that isocontour of the objective function with minimum value, with at least one point of intersection with the feasible region.

The geometrical approach to mathematical programming problems in later sections will bring out more of the differences between linear and nonlinear programming problems.

Quadratic Programming

Define

$$B = (b_{jk}), \quad j = 1, \ldots, n \text{ and } k = 1, \ldots, n,$$

a symmetric positive semidefinite matrix. [This means that for every vector $z = (z_1, \ldots, z_n)^t$, $z^t B z \geq 0$.] The quadratic programming problem is to choose $x = (x_1, \ldots, x_n)^t$ to

$$\text{minimize } c^t x + x^t B x \tag{1.6}$$

subject to

$$Ax - b \geq 0, \quad x \geq 0. \tag{1.7}$$

The problem is characterized by linear constraints, non-negativity requirements on the variables, and an objective function that is in a positive semidefinite form. Unlike the linear programming problem, the solution to a quadratic problem need not lie on the vertex of the convex polyhedron formed by the constraints (1.7), as the following example shows. The problem is to choose x_1 and x_2 to

$$\text{minimize } x_1^2 - x_1 + x_2^2 - \frac{x_2}{2}$$

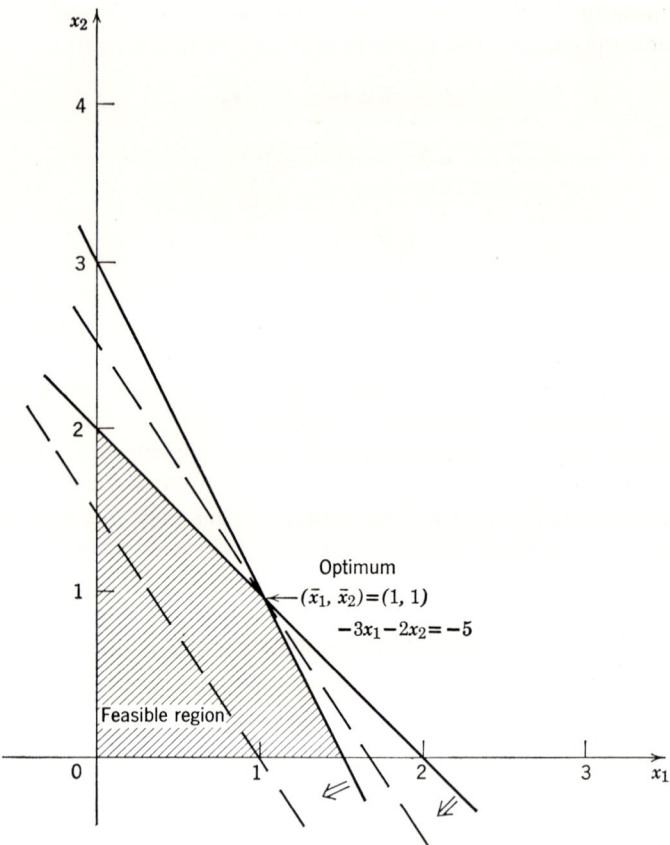

Minimize
$$-3x_1 - 2x_2$$
subject to
$$\begin{aligned} -2x_1 - x_2 + 3 &\geq 0 \\ -x_1 - x_2 + 2 &\geq 0 \\ x_1 &\geq 0 \\ x_2 &\geq 0 \end{aligned}$$

Figure 1.1 Linear Programming Problem

subject to
$$-x_1 - x_2 + 1 \geq 0,$$
$$x_1 \geq 0, \qquad x_2 \geq 0.$$

Here
$$B = \begin{bmatrix} 1 & 0 \\ 0 & 1 \end{bmatrix}, \quad c = (-1, -\tfrac{1}{2})^t, \quad A = [-1, -1], \quad b = -1$$

Structure of Mathematical Programming Problems

Since

$$[z_1, z_2] B \begin{bmatrix} z_1 \\ z_2 \end{bmatrix} = z_1^2 + z_2^2 \geq 0 \quad \text{for all} \quad (z_1, z_2),$$

B satisfies the definition of a positive semidefinite matrix.

This problem is presented graphically in Figure 1.2. The shaded region is the set of points satisfying $x_1 \geq 0$, $x_2 \geq 0$, and $-x_1 - x_2 + 1 \geq 0$. The isocontours of the objective function are circles, something that never happens in linear programming. They represent decreasing values of the

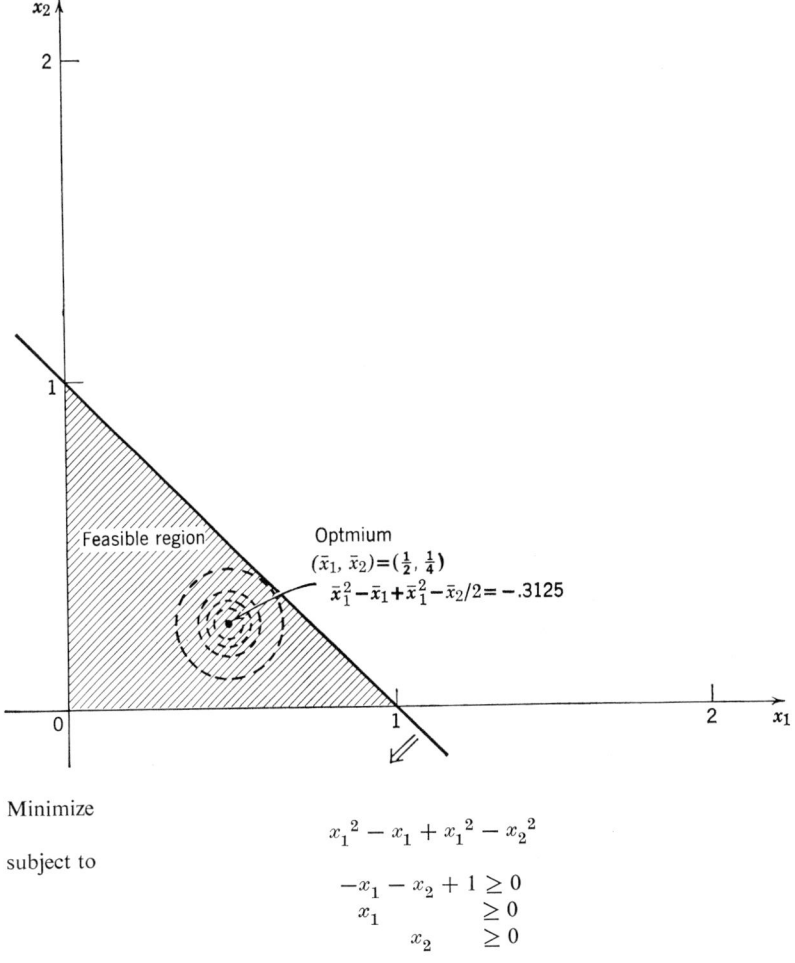

Minimize
$$x_1^2 - x_1 + x_1^2 - x_2^2$$

subject to
$$-x_1 - x_2 + 1 \geq 0$$
$$x_1 \geq 0$$
$$x_2 \geq 0$$

Figure 1.2 Quadratic Programming Problem

objective function as they move into $(\frac{1}{2}, \frac{1}{4})$, which is the solution to the example. This point does not lie on the boundary of the shaded (feasible) region. This possibility is one central difference between linear and nonlinear programming problems, a particular case of which is illustrated here by this quadratic programming problem.

By using known theoretical properties of the solution of a quadratic programming problem it is possible to develop special-purpose algorithms that solve them in a *finite* number of steps, usually by some variation of the simplex method for linear programming. For references, see Hadley [24], pp. 240 and 241.

A great deal of work has been done in the area of solving quadratic programming problems with some emphasis on special structure—for instance, efficient ways to handle upper and lower bounds on the variables. This effort seems particularly out of proportion when one considers the class of models that fit the special requirements of the quadratic programming problem. There are no quadratic programming models in this book. Two quadratic programming problems in the literature are the portfolio selection problem of Markowitz [30], in which the objective function (to be minimized) is the variance of a probability distribution, a well-known positive semi-definite form, and the problem of least-squares estimation of the parameters of a linear regression model with constraints on the parameters and/or deviations. There is another possible use of these algorithms, as sub-algorithms for solving nonlinear problems whose objective functions can be locally approximated by a quadratic positive semidefinite form, and whose constraints can be linearly approximated locally.

Local and Global Minima

The reason why B must be a positive semidefinite matrix in the quadratic programming problem provides an introduction to the most serious problem found in nonlinear programming: the problem of local and global minima.

A *local minimum* is a point (x_L) in the feasible region about which all indications are that x_L is the solution to the problem at hand. The *global minimum* to a problem is the local minimum of all possible local minima with the smallest value of the objective function. When the nonlinear programming problem has certain characteristics, such as in the quadratic programming problem discussed above, that is, when B is a positive semidefinite matrix, it can be shown that any local minimum is also a global minimum. When B is not a positive semidefinite matrix, there is a possibility that several local minima exist.

Consider the following example. The problem is to choose x_1 and x_2 to

$$\text{minimize } -x_1^2 - x_2^2$$

Structure of Mathematical Programming Problems

subject to
$$-x_1 - 4x_2 + 5 \geq 0,$$
$$-x_1 + 1 \geq 0,$$
$$x_1 \geq 0, \quad x_2 \geq 0.$$

The matrix $B = \begin{bmatrix} -1 & 0 \\ 0 & -1 \end{bmatrix}$, which is not a positive semidefinite matrix.

As shown in Figure 1.3, the isocontours of the objective function are arcs of circles with the origin as center. The values of the isocontours decrease in any direction away from the origin. As is seen by inspection, the points $(0, 1\frac{1}{4})$ and $(1, 1)$ are both local minima to the problem. Any attempt to leave $(0, 1\frac{1}{4})$ must of necessity leave the region of feasibility or increase the value of the objective function. In short, all the information about that point indicates that it is the solution to the problem. The same analysis applies to $(1, 1)$, and thus it is also a local minimum. Obviously, because $f(1, 1) < f(0, 1\frac{1}{4})$, $(1, 1)$ is also the global minimum. In a large problem one does not know whether all the local minima have been found, and thus no algorithm can guarantee convergence to the global solution.

Mathematical researchers, in order to satisfy the desire for proof of convergence to the global optimum, have been influenced to consider mathematical programming problems (as the quadratic one stated at the beginning of this section) where this could be done.

Convex Programming

The problem of choosing x to

$$\text{minimize } f(x) \quad (f \text{ a convex function}) \tag{1.8}$$

subject to

$$g_i(x) \geq 0, \quad i = 1, \ldots, m \quad (g_i \text{ a concave function for all } i) \tag{1.9}$$

is called the convex programming problem. Definitions and examples are given below.

The same mathematical considerations that lead to the quadratic programming format, which avoids the local-global minimum problem, also lead to this general class of problems for which any local solution is a global solution. In an important paper in 1951 [27] Kuhn and Tucker established necessary and sufficient conditions for a point (vector) to solve the convex programming problem defined by (1.8) and (1.9).

The term convex is applied to sets of points and to functions. In the first case, for T to be a *convex set of points*, every straight line connecting any two points in T must be contained entirely in T. A function f is a *convex function* of x if it is never underestimated by linear interpolation, or, for

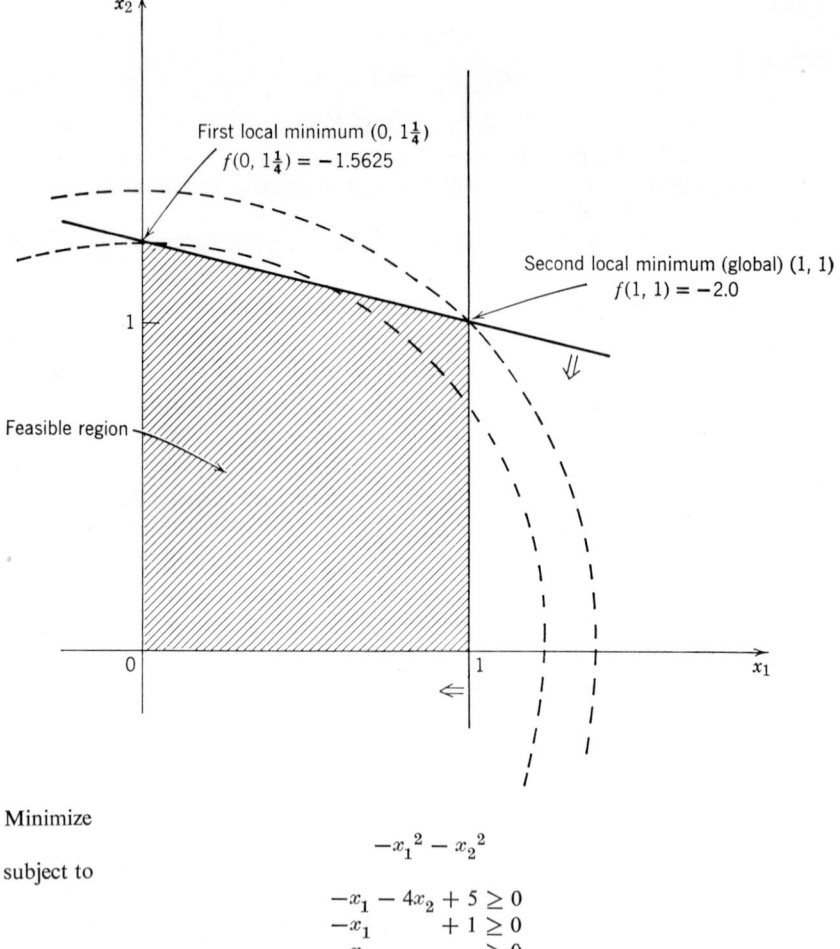

Minimize
$$-x_1^2 - x_2^2$$
subject to
$$-x_1 - 4x_2 + 5 \geq 0$$
$$-x_1 + 1 \geq 0$$
$$x_1 \geq 0$$
$$ x_2 \geq 0$$

Figure 1.3 Quadratic Programming Problem with Two Local Minima

every scalar λ, where $0 < \lambda < 1$, and any two vectors $x^{(1)}$ and $x^{(2)}$,

$$f[\lambda x^{(1)} + (1 - \lambda)x^{(2)}] \leq \lambda f(x^{(1)}) + (1 - \lambda)f(x^{(2)}). \qquad (1.10)$$

The two notions of convexity of sets of points and of functions are connected in the following way. If $f(x)$ is a convex function, then for any number k the set of points x for which $f(x) \leq k$ is a convex set. A concave function is one whose negative is a convex function. Thus the convex programming problem is sometimes specified in terms of maximizing $h(x)$, where $h(x)$ is a

Structure of Mathematical Programming Problems

concave function. It follows from the above that, if $g(x)$ is a concave function, then for any number k the set of points for which $g(x) \geq k$ is a convex set.

Many other properties follow from the definitions of convex and concave. In particular it should be noted that the points common to two or more convex sets form a convex set. Also, the constraints to the convex programming problem are sometimes written $h_i(x) \leq 0, i = 1, \ldots, m$, where each $h_i(x)$ is a convex function.

The convex programming problem is the problem of minimizing a convex function (or equivalently maximizing a concave function) in a convex set of points.

Consider the example of choosing x_1 and x_2 to

$$\text{minimize } f(x_1, x_2) = (x_1 - 2)^2 + (x_2 - 1)^2$$

subject to

$$g_1(x_1, x_2) = -x_1^2 + x_2 \geq 0$$
$$g_2(x_1, x_2) = -x_1 - x_2 + 2 \geq 0.$$

The geometry of the example is given in Figure 1.4. It can be verified that $f(x_1, x_2)$ is convex and that $g_1(x_1, x_2)$ is concave. The function $g_2(x_1, x_2)$ is a linear function. Linear functions are always both convex *and* concave. From the figure it is easy to verify that each constraint defines a convex set, and the points satisfying both constraints (the shaded feasible region) form a convex set. The points yielding smaller objective function values than those on any particular isocontour form a convex set, a direct property of the definition of a convex function.

A general proof can be given that any local solution to the convex programming problem is a global solution, and this example indicates geometrically how to go about proving it.

Since the appearance of the Kuhn-Tucker paper [27], many algorithms have been directed toward solving the convex programming problem.

One method that would seem useful is to convert the problem to one of minimizing a linear objective function while approximating the boundary of the feasible region by a convex polyhedron. This suggestion (roughly) was made independently by Cheney and Goldstein [12] and Kelley [26], and is called "the method of cutting planes." This method relies almost exclusively on the fact that the tangent plane to any point on the boundary of a convex region lies entirely outside the region. For this reason it is not well suited to any class of problems other than convex ones. Even for this class, computational experience using the method of cutting planes has not been extensive.

Cutting plane methods are conceptually related to generalized linear programming methods (decomposition methods), which attempt to convert nonlinear programming problems into linear programming problems by approximating the convex set by chords drawn between extreme points [15].

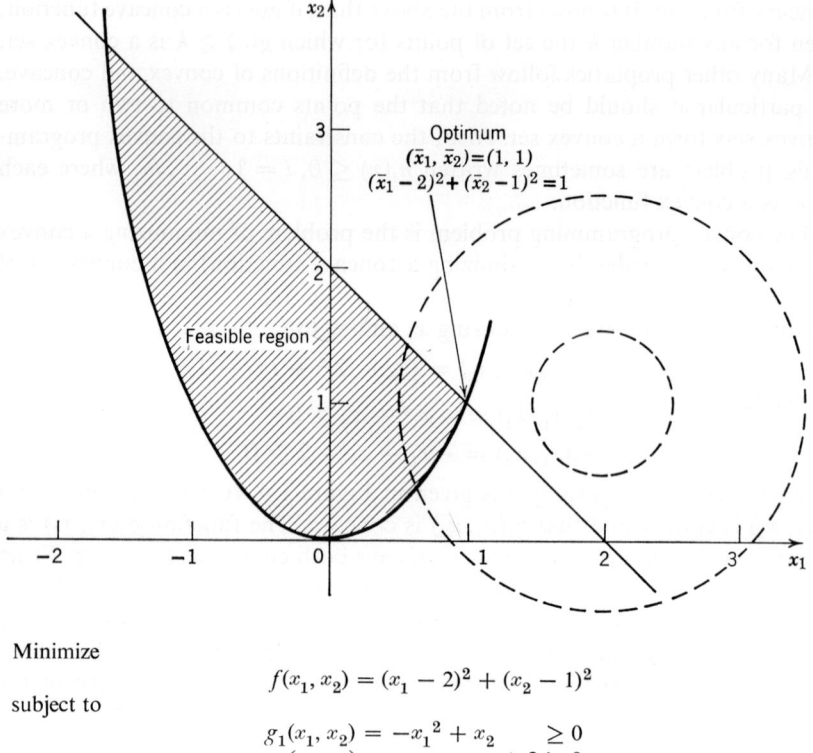

Minimize
$$f(x_1, x_2) = (x_1 - 2)^2 + (x_2 - 1)^2$$
subject to
$$g_1(x_1, x_2) = -x_1^2 + x_2 \geq 0$$
$$g_2(x_1, x_2) = -x_1 - x_2 + 2 \geq 0$$

Figure 1.4 Convex Programming Problem

A second approach developed to solve convex programming problems can be classified as methods of feasible directions. Three particularly important algorithms are those of Zoutendijk [47], Rosen's gradient projection techniques [35, 36], and the reduced gradient method [1, 46]. These algorithms have been extensively tested computationally for nonlinear objective function and linear constraints, but, because of the many difficulties inherent in solving problems with nonlinear objective functions and nonlinear constraints, computational verification of the algorithms has been quite limited. For additional references and discussion of these methods, see Chapter 9 of Hadley [24].

A third and quite recently developed approach, extensively tested for nonlinear objective function and linear and/or nonlinear constraints, is described in Section 1.4. It is the sequential unconstrained minimization technique (SUMT) of Fiacco and McCormick, and has been used to solve the problems described in this book.

Structure of Mathematical Programming Problems

Several of the examples in this book are convex programming problems. The weapons assignment problem is a convex programming problem because the constraints are linear and the objective function (to be maximized) is concave. The bid evaluation problem has linear constraints and a separable piecewise linear objective function. The general characteristic of the objective function, excluding the origin, where it may not be continuous, is that of a *concave* function. Quite often objective functions representing total cost are concave and hence pose mathematical difficulties in determining global solutions. The cattle feed problem has a linear objective function, one concave ($\geq$) constraint, and the remainder linear constraints. The stratified sampling problem has a linear objective function and several convex ($\leq$) constraints. The chemical equilibrium problem has a convex objective function and linear constraints.

The alkylation problem has many badly behaved (nonconvex or nonconcave) functions: the objective function, the constraints representing bounds on the fit to experimental data, and the nonlinear equality constraints.

The classic linear regression problem is the minimization of a convex function, but any curve-fitting problem where the parameters enter nonlinearly is a nonconvex programming problem, as in the example presented in Chapter 8. Similarly, in general a maximum likelihood estimation is the maximization of a nonconcave function.

Separable Programming

The problem of choosing x_j ($j = 1, \ldots, n$) to

$$\text{minimize } f(x) = \sum_{j=1}^{n} f_j(x_j) \tag{1.11}$$

subject to

$$g_i(x) = \sum_{j=1}^{n} g_{ij}(x_j) \geq 0, \quad i = 1, \ldots, m \tag{1.12}$$

is called a *separable* programming problem. Essentially this means that variables are not coupled together in any of the problem functions except in an additive fashion. Mathematically it can be expressed (for all x_j) as

$$\frac{\partial^2 f(x)}{\partial x_k \, \partial x_j} = 0 \quad \text{for} \quad k \neq j, \quad \frac{\partial^2 g_i}{\partial x_k \, \partial x_j} = 0 \quad \text{for} \quad k \neq j, \quad i = 1, \ldots, m. \tag{1.13}$$

The notion of a separable function is not directly related to the concepts discussed thus far. All linear programming problems are separable. The quadratic programming problem is separable if and only if B is a diagonal matrix. (Note that the linear programming example and the two quadratic programming examples were separable.)

Sometimes separability has the more general meaning that any problem is separable if it can be reduced to the form (1.11) and (1.12) by a linear transformation of the variables. Using this extended notion the quadratic programming problem is separable. Since the objective function of the weapons assignment problem can be made to separate by a linear transformation, it also is a separable programming problem. This extended notion of separability is a very powerful one. Many problems fall into this form regardless of convexity or concavity. Because of this, many algorithms have been devised to solve just this kind of problem. For more information see [11] and Chapter 10 of [8].

The bid evaluation problem is separable. The alkylation problem cannot be separated by any linear transformation. The chemical equilibrium problem is not separable, nor are any of the curve-fitting problems or the cattle feed problem. The stratified sampling problem is given in separable form.

1.4 THE SEQUENTIAL UNCONSTRAINED MINIMIZATION TECHNIQUE FOR NONLINEAR PROGRAMMING

Minimizing a nonlinear function subject to linear constraints is much less difficult then when the constraints are nonlinear. The main reason is that it is very difficult to move along the boundary of a nonlinearly constrained region, whereas it is relatively simple to move along the boundary of a linearly constrained region. Rosen's projected gradient method for nonlinear programming with linear constraints [35] provides a good procedure for moving along the boundary of linearly constrained regions. In order to solve the nonlinearly constrained problem, an idea for handling nonlinear constraints was proposed by C. W. Carroll [4, 5]. The mathematical validity and computational implementation have been developed by Fiacco and McCormick [17–20].

It is convenient to write the nonlinear programming problem with nonlinear inequality and equality constraints in the following ways. Choose x to

$$\text{minimize } f(x) \tag{1.14}$$

subject to

$$g_i(x) \geq 0, \quad i = 1, \ldots, l, \tag{1.15}$$

$$g_i(x) = 0, \quad i = l+1, \ldots, m, \tag{1.16}$$

where there exists at least one point x such that $g_i(x) > 0$, for $i = 1, \ldots, l$. To solve this problem the following algorithm is proposed. Define the function (called the P function)

$$P(x, r_1) \equiv f(x) + r_1 \sum_{i=1}^{l} \frac{1}{g_i(x)} + r_1^{-\frac{1}{2}} \sum_{i=l+1}^{m} g_i^2(x), \tag{1.17}$$

where r_1 is a positive number.

Sequential Unconstrained Minimization Technique

As a starting point determine x^0 such that $g_i(x^0) > 0, i = 1, \ldots, l$. Proceed from x^0 to a point $x(r_1)$ that approximates the minimum of $P(x, r_1)$ in the set of points satisfying (1.15). Form the new function

$$P(x, r_2) \equiv f(x) + r_2 \sum_{i=1}^{l} \frac{1}{g_i(x)} + r_2^{-\frac{1}{2}} \sum_{i=l+1}^{m} g_i^2(x), \tag{1.18}$$

where $r_1 > r_2 > 0$. Starting from $x(r_1)$, approximate the minimum of $P(x, r_2)$.

By continuing in this manner, a sequence of points $\{x(r_k)\}, k = 1, 2, 3, \ldots$ is generated that respectively minimize $\{P(x, r_k)\}$, where $\{r_k\}$ is a sequence that decreases strictly monotonically to 0. The conjecture (which must be proved) is that the sequence of unconstrained minima $\{x(r_k)\}$ will approach a solution to the mathematical programming problem defined by (1.14), (1.15), and (1.16) as r_k goes to 0. The intuitive reasons for this can be summarized as follows.

The term $r_k \sum_{i=1}^{l} 1/g_i(x)$ is regarded as a "penalty" factor attached to the objective function $f(x)$ and assures that a minimum to the P function is achieved in the interior of the inequality-constrained region by balancing the avoidance of boundaries and minimization of $f(x)$. This can be seen intuitively. Consider the trajectory of points that tend to minimize $P(x, r_1)$, starting from x^0. By assumption, $g_i(x^0) > 0$, all i, and so $P(x, r_1)$ exists and has some finite value. Since this trajectory defines a curve on which P is continually decreasing, no point on the trajectory can yield a P function value exceeding $P(x^0, r_1)$. Since the feasible boundary is defined by one or more of the $g_i(x) = 0$, it is apparent that $P \to +\infty$ as the boundary is approached from any interior point. Consequently the boundary can never be pierced by the trajectory and the minimum of $P(x, r_1)$ must be a feasible interior point.

The intuitive motivation for the third term is clear. As $r_k \to 0$, the third term would tend to $+\infty$ unless each $g_i[x(r_k)]$ went to zero. Thus minimizing the P function would tend to force the g_i's to zero.

Another motivation behind this formulation is the transformation of the original constrained problem into a sequence of unconstrained minimization problems. The desirability of this lies in the fact that a number of methods for minimizing an unconstrained function are known and many newer ones are being developed [16, 21, 22, 33, 34, 41]. Thus by this transformation it becomes possible to solve the more formidable constrained problem without inventing new techniques.

A very desirable feature of this transformation with respect to a problem previously mentioned is that it avoids the necessity of coping separately with the boundary of the inequality-constrained region, for example, by attempting to move along the boundary once it is encountered. Such motion is cumbersome when the constraining surface is nonlinear. The P function treats the objective function in a manner that makes it possible to eliminate motion along the boundary.

The theoretical validity of the well-behaved (convex) problem is given in [18, 20]. Under the convexity conditions this approach is *guaranteed* to solve the programming problem. For nonconvex problems Stong [43] proved the existence of a sequence of global P function minima converging to the global solution of the programming problem.

Research topics currently include analysis of problem structure in order to solve larger problems, exploration of methods of minimizing nonconvex functions, extrapolation procedures for accelerating convergence, and investigation of combining SUMT with simplicial methods to yield more efficient algorithms.

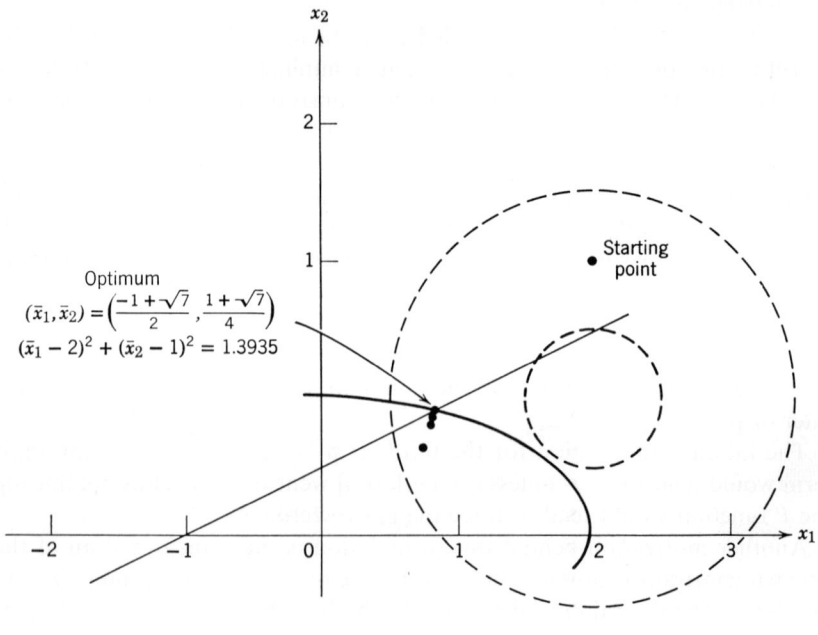

Minimize

$$f(x_1, x_2) = (x_1 - 2)^2 + (x_2 - 1)^2$$

subject to

$$g_1(x_1, x_2) = -\frac{x_1^2}{4} - x_2 + 1 \geq 0$$

$$g_2(x_1, x_2) = x_1 - 2x_2 + 1 \geq 0$$

Figure 1.5 SUMT Solution of Example Problem

Table 1.1 Data from SUMT Iterations

r	$x_1(r)$	$x_2(r)$	$f(r)$
1.0	.7489	.5485	1.7691
4.0×10^{-2}	.8177	.8323	1.4258
1.6×10^{-3}	.8224	.8954	1.3976
6.4×10^{-5}	.8228	.9082	1.3942
2.56×10^{-6}	.8229	.9108	1.3936
1.024×10^{-7}	.8229	.9113	1.3935
4.096×10^{-9}	.8229	.9114	1.3935

Starting point: $(x_1, x_2) = (2.0, 2.0)$
Theoretical solution: $(\bar{x}_1, \bar{x}_2) = (.8229, .9114), \bar{f} = 1.3935$

Consider the following example. Choose x_1 and x_2 to

$$\text{minimize } f(x_1, x_2) = (x_1 - 2)^2 + (x_2 - 1)^2$$

subject to

$$g_1(x_1, x_2) = -\frac{x_1^2}{4} - x_2^2 + 1 \geq 0,$$

$$g_2(x_1, x_2) = x_1 - 2x_2 + 1 = 0.$$

The optimum as seen from Figure 1.5 is at $[(-1 + \sqrt{7})/2 \simeq .8229, (1 + \sqrt{7})/4 \simeq .9114]$, where the value of the objective function is $9 - (23/8)\sqrt{7} \simeq 1.3935$. In Table 1.1 the solutions to this problem, as generated by the computer program implementing SUMT [31], are given. The trajectory generated is also plotted in Figure 1.5. Note that the boundary of the ellipse defined by $g_1(x_1, x_2)$ is approached as r_k goes to 0 and that the equality constraint is also satisfied in the limit.

References

[1] J. Abadie, J. Carpentier, and C. Hensgen, "Generalization of the Wolfe Gradient Method to the Case of Nonlinear Constraints," paper presented at the Joint European Meeting of the Econometric Society/The Institute of Management Science, Warsaw, September 1966.
[2] R. Bellman, *Dynamic Programming*, Princeton University Press, Princeton, N.J., 1957.
[3] R. Bellman and S. Dreyfus, *Applied Dynamic Programming*, Princeton University Press, Princeton, N.J., 1962.
[4] C. W. Carroll, "An Operations Research Approach to the Economic Optimization of a Kraft Pulping Process," Ph.D. Dissertation, Institute of Paper Chemistry, Appleton, Wis., 1959.
[5] C. W. Carroll, "The Created Response Surface Technique for Optimizing Nonlinear Restrained Systems," *Operations Res.*, **9**, 169–184 (1961).
[6] A. Cauchy, "Méthode Générale pour la Resolution des Systèmes D'équations Simultanées, *Compt. Rend.*, **25**, 536–538 (1847).

[7] A. Charnes and W. W. Cooper, "Chance-Constrained Programming," *Management Sci.*, **6**, 73–79 (1959).
[8] A. Charnes and W. W. Cooper, *Management Models and Industrial Applications of Linear Programming*, Vols. I and II, Wiley, New York, 1961.
[9] A. Charnes and W. W. Cooper, "Chance Constraints and Normal Deviates," *J. Am. Statist. Assoc.*, **57**, 134–148 (1962).
[10] A. Charnes and M. J. L. Kirby, "Optimal Decision Rules for the E Model of Chance-Constrained Programming," Technical Paper TP-166, Research Analysis Corporation, McLean, Va., July 1965.
[11] A. Charnes and C. E. Lemke, "Minimization of Nonlinear Separable Convex Functions," *Naval Res. Logist. Quart.*, **6**, 43–46 (1959).
[12] E. W. Cheney and A. A. Goldstein, "Newton's Method for Convex Programming and Tchebycheff Approximation," *Numer. Math.*, **1**, 253–268 (1959).
[13] R. Courant and D. Hilbert, *Methods of Mathematical Physics*, Vol. 1, Interscience, Publishers, New York, 1953.
[14] G. B. Dantzig, "Maximization of a Linear Function of Variables Subject to Linear Inequalities," in T. C. Koopmans (Ed.), *Activity Analysis of Production and Allocation*, Wiley, New York, 1951.
[15] G. B. Dantzig, *Linear Programming and Extensions*, Princeton University Press, Princeton, N.J., 1963.
[16] W. C. Davidon, "Variable Metric Method for Minimization," Research and Development Report ANL-5990 (Revised), Argonne National Laboratory, U.S. Atomic Energy Commission, 1959.
[17] A. V. Fiacco and G. P. McCormick, "Programming under Nonlinear Constraints by Unconstrained Minimization: A Primal-Dual Method," Technical Paper RAC-TP-96, Research Analysis Corporation, McLean, Va., September 1963.
[18] A. V. Fiacco and G. P. McCormick, "The Sequential Unconstrained Minimization Technique for Nonlinear Programming: A Primal-Dual Method," *Management Sci.*, **10**, 360–366 (1964).
[19] A. V. Fiacco and G. P. McCormick, "Computational Algorithm for the Sequential Unconstrained Minimization Technique for Nonlinear Programming," *Management Sci.*, **10**, 601–617 (1964).
[20] A. V. Fiacco and G. P. McCormick, "Extensions of SUMT for Nonlinear Programming: Equality Constraints and Extrapolation," *Management Sci.*, **12**, 816–828 (1966).
[21] R. Fletcher and M. J. D. Powell, "A Rapidly Convergent Descent Method for Minimization," *Computer J.*, **6**, 163–168 (1963).
[22] R. Fletcher and C. M. Reeves, "Function Minimization by Conjugate Gradients," *Computer J.*, **7**, 149–154 (1964).
[23] D. Gale, H. W. Kuhn, and A. W. Tucker, "Linear Programming and the Theory of Games," in T. C. Koopmans (Ed.), *Activity Analysis of Production and Allocation*, Wiley, New York, 1951.
[24] G. Hadley, *Nonlinear and Dynamic Programming*, Addison-Wesley, Reading, Mass., 1964.
[25] R. R. Hughes, E. Singer, and M. Souders, "Machine Design of Refineries," *Proc. World Petroleum Conf.*, Sixth, Frankfurt-am-Main (June 19–26, 1963).
[26] J. E. Kelley, Jr., "The Cutting Plane Method for Solving Convex Programs," *J. Soc. Ind. Appl. Math.*, **8**, 703–712 (1960).
[27] H. W. Kuhn and A. W. Tucker, "Nonlinear Programming," in J. Neyman (Ed.), *Proceedings of the Second Berkeley Symposium on Mathematical Statistics and Probability*, University of California Press, Berkeley, 1951, pp. 481–493.

References

[28] C. E. Lemke, "Bimatrix Equilibrium Points and Mathematical Programming," *Management Sci.*, **11**, 681–689 (1965).

[29] O. L. Mangasarian and H. Stone, Two-Person Nonzero-Sum Games and Quadratic Programming, *J. Math. Anal. Appl.*, **9**, 348–355 (1964).

[30] H. Markowitz, *Portfolio Selection*, Wiley, New York, 1959.

[31] G. P. McCormick, W. C. Mylander, III, and A. V. Fiacco, "Computer Program Implementing the Sequential Unconstrained Minimization Technique for Nonlinear Programming," Technical Paper TP-151, Research Analysis Corporation, McLean, Va., April 1965 (see also SHARE Distribution No. SDA 3189).

[32] L. W. Neustadt, "Optimization, a Moment Problem, and Nonlinear Programming," *J. Soc. Ind. Appl. Math. Control, Ser. A*, **2**, 33–53 (1964).

[33] M. J. D. Powell, "An Efficient Method of Finding the Minimum of a Function of Several Variables without Calculating Derivatives," *Computer J.*, **7**, 155–162 (1964).

[34] M. J. D. Powell, "Finding Minima of Functions of Several Variables," paper presented at the Institute of Mathematics and its Applications Conference, Birmingham University, July 1965 (to be published by Academic Press, London).

[35] J. B. Rosen, "The Gradient Projection Method for Nonlinear Programming, Part I: Linear Constraints," *J. Soc. Ind. Appl. Math.*, **9**, 181–217 (1960).

[36] J. B. Rosen, "The Gradient Projection Method for Nonlinear Programming, Part II: Nonlinear Constraints," *J. Soc. Ind. Appl. Math.*, **9**, 514–532 (1961).

[37] J. B. Rosen, "Existence and Uniqueness of Equilibrium Points for Concave, N-Person Games," *Econometrica*, **33**, 520–534 (1965).

[38] J. B. Rosen, "Iterative Solution of Nonlinear Optimal Control Problems," Technical Summary Report 591, U.S. Army Mathematics Research Center, Madison, Wis. July 1965.

[39] H. Rubin and P. Unger, "Motion Under a Strong Constraining Force," *Commun. Pure Appl. Math.*, **10**, 65–87 (1957).

[40] L. A. Schmidt, Jr., and R. L. Fox, "An Integrated Approach to Structural Synthesis and Analysis," paper presented at the AIAA Fifth Annual Structures and Materials Conference, Palm Springs, Calif., April 1–3, 1964.

[41] H. A. Spang, III, "A Review of Minimization Techniques for Nonlinear Functions," *SIAM. Rev.*, **4**, 343–365 (1962).

[42] L. G. Stoleru, "An Optimal Policy for Economic Growth," unpublished paper, Institute for Mathematical Studies in the Social Sciences, Stanford University, Palo Alto, Calif., 1964.

[43] R. E. Stong, "A Note on the Sequential Unconstrained Minimization Technique for Nonlinear Programming," *Management Sci.*, **12**, 142–144 (1965).

[44] S. H. Walker, "Estimation of the Probability of Occurrences of a Disease as a Function of Several Variables," Ph.D. dissertation, Johns Hopkins University, Baltimore, May 1965.

[45] P. Wolfe, "A Duality Theorem for Nonlinear Programming," *Quart. Appl. Math.*, **19**, 239–244 (1961).

[46] P. Wolfe, "Methods of Nonlinear Programming," in R. Graves and P. Wolfe (Eds.), *Recent Advances in Mathematical Programming*, McGraw-Hill, New York, 1963, pp. 67–86.

[47] G. Zoutendijk, *Methods of Feasible Directions*, Elsevier, Amsterdam, 1960.

2
WEAPONS ASSIGNMENT

In this chapter we discuss two weapons assignment models. The first model determines an assignment of weapons to targets to maximize expected target damage value. There are constraints on weapons available of various types and on the minimum number of weapons by type to be assigned to various targets. The constraints are linear, and the objective function is nonlinear. The second model determines an assignment of weapons to targets such that weapons cost is minimized and at least a specified expected damage value is inflicted both on various targets and on various target classes. The constraints are nonlinear, and the objective function is linear.

2.1 MODEL FOR MAXIMIZING EXPECTED TARGET DAMAGE VALUE

The variable to be determined is

x_{ij} = the number of weapons of type i assigned to target j,

$$i = 1, \ldots, p \quad \text{and} \quad j = 1, \ldots, q.$$

Limitations on the number of weapons assigned are specified in terms of

a_i = the total number of weapons of type i available,

b_j = the minimum number of weapons of all types assigned to target j.

The constraints on total number of weapons and on minimum weapons assigned to targets are

$$\sum_{j=1}^{q} x_{ij} \leq a_i, \quad i = 1, \ldots, p, \tag{2.1}$$

$$\sum_{i=1}^{p} x_{ij} \geq b_j, \quad j = 1, \ldots, q. \tag{2.2}$$

The objective function is formulated in terms of probability of damage of various targets weighted by their military value. Define

α_{ij} = the probability that target j will be undamaged by an attack using one unit of weapon i

and

u_j = the military value of target j.

The expected damage to target j by an assignment of x_{ij} weapons of type i is $[1 - \alpha_{ij}^{x_{ij}}]$, and the expected damage to target j by the over-all assignment of weapons of all types $\sum_{i=1}^{p} x_{ij}$ is $[1 - \prod_{i=1}^{p} \alpha_{ij}^{x_{ij}}]$. The total expected target damage value is the sum of the expected damage to targets weighted by the military value of the targets,

$$\sum_{j=1}^{q} u_j \left[1 - \prod_{i=1}^{p} \alpha_{ij}^{x_{ij}} \right]. \quad (2.3)$$

The nonlinear programming problem is as follows. Choose x_{ij}'s to maximize the nonlinear objective function (2.3) subject to linear constraints (2.1) and (2.2) and to the non-negativity conditions

$$x_{ij} \geq 0, \quad i = 1, \ldots, p \quad \text{and} \quad j = 1, \ldots, q. \quad (2.4)$$

An alternative definition of α_{ij} is the fraction of target j that will not be damaged by an attack using one unit of weapon i. The objective function is then interpreted as the total fractional damage value (to be maximized).

2.2 MODEL FOR INFLICTING SPECIFIED DAMAGE WITH MINIMUM COST

The variable to be determined is x_{ij} ($i = 1, \ldots, p$ and $j = 1, \ldots, q$), the number of weapons of type i assigned to target j, as defined previously. The probability α_{ij} that target j will be undamaged by an attack using one unit of weapon i is also defined as before.

Constraints on the amount of damage to be inflicted on the targets are specified in terms of

d_j = the minimum expected damage to target j,
$d^{(k)}$ = the minimum expected weighted damage to targets of class k,

where $k = 1, \ldots, r$,

$v_j^{(k)}$ = the weight of target j with respect to the targets of class k.

Constraints on expected damage to the various targets may be written

$$1 - \prod_{i=1}^{p} \alpha_{ij}^{x_{ij}} \geq d_j, \quad j = 1, \ldots, q. \quad (2.5)$$

Constraints on weighted expected damage to the various target classes may be written

$$\sum_{j=1}^{q} v_j^{(k)} \left[1 - \prod_{i=1}^{p} \alpha_{ij}^{x_{ij}} \right] \geq d^{(k)}, \qquad k = 1, \ldots, r. \tag{2.6}$$

For purposes of formulating the objective function, let us assume that the cost of weapons assignment is linear over the range being considered, and define

$$c_i = \text{the cost per unit of weapon of type } i.$$

The total cost of the assignment of x_{ij} ($i = 1, \ldots, p$ and $j = 1, \ldots, q$) weapons to the various targets is

$$\sum_{i=1}^{p} c_i \sum_{j=1}^{q} x_{ij}. \tag{2.7}$$

The nonlinear programming problem is as follows. Choose x_{ij}'s to minimize the linear objective function (2.7) subject to nonlinear constraints (2.5) and (2.6) and to non-negativity restrictions (2.4).

It should be noted that constraint (2.5) can be converted into a linear constraint by appropriate logarithmic transformation.

An alternative definition of α_{ij} would again be the fraction of target j undamaged by an attack using one unit of weapon i. The damage constraints may be interpreted as requiring certain fractional damage values to various targets.

2.3 EXAMPLE

As an example we consider a model for maximizing expected target damage value such as was given in Section 2.1. Weapons of five types are to be assigned to 20 different targets. Upper limits on available weapons and lower limits on weapons to be assigned are specified.

Figure 2.1 is a map of a hypothetical area containing the 20 targets, and on which the range of the five weapon types is indicated. The characteristics of the five weapon types could be thought of as follows:

1. Intercontinental ballistic missiles.
2. Medium-range ballistic missiles from first firing area.
3. Long-range bombers.
4. Fighter bombers.
5. Medium-range ballistic missiles from second firing area.

Table 2.1 gives the values of the parameters needed for the model: probabilities that targets will be undamaged by weapons, total number of weapons available, minimum number of weapons to be assigned, and military value of targets.

Example

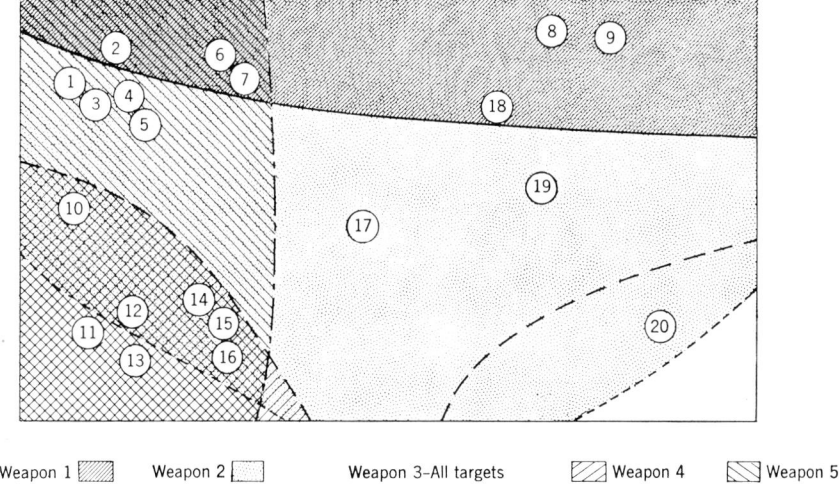

Figure 2.1 Map of Targets and Coverage of Weapons

Using the information given in Table 2.1, we formulate the model for maximizing expected target damage value as follows. The criterion function to be maximized is the total expected target damage:

$$60[1.00 - (1.00^{x_{11}} \cdot .84^{x_{21}} \cdot .96^{x_{31}} \cdot 1.00^{x_{41}} \cdot .92^{x_{51}})]$$
$$+ \cdots$$
$$+ 150[1.00 - (1.00^{x_{1,20}} \cdot .85^{x_{2,20}} \cdot .92^{x_{3,20}} \cdot 1.00^{x_{4,20}} \cdot 1.00^{x_{5,20}})].$$

The linear constraints on the total number of weapons of the five types are

$$x_{11} + \cdots + x_{1,20} \leq 200$$
$$\vdots \qquad \qquad \vdots$$
$$x_{51} + \cdots + x_{5,20} \leq 250,$$

and the linear constraints on the minimum assignment of weapons to the seven specified targets that must be attacked are

$$x_{11} + \cdots + x_{51} \geq 30$$
$$x_{16} + \cdots + x_{56} \geq 100$$
$$\vdots \qquad \qquad \vdots$$
$$x_{1,20} + \cdots + x_{5,20} \geq 10.$$

Table 2.1 Values of Parameters for Example

										α_{ij}: Probability That Weapon i Will Not Damage Target j											
										Targets											a_i: No. of Weapons i Available
Weapon	1	2	3	4	5	6	7	8	9	10	11	12	13	14	15	16	17	18	19	20	
1	1.00	.95	1.00	1.00	1.00	.85	.90	.85	.80	1.00	1.00	1.00	1.00	1.00	1.00	1.00	1.00	.95	1.00	1.00	200
2	.84	.83	.85	.84	.85	.81	.81	.82	.80	.86	1.00	.98	1.00	.88	.87	.88	.85	.84	.85	.85	100
3	.96	.95	.96	.96	.96	.90	.92	.91	.92	.95	.91	.98	.99	.98	.97	.98	.95	.92	.93	.92	300
4	1.00	1.00	1.00	1.00	1.00	1.00	1.00	1.00	1.00	.96	.91	.92	.91	.92	.98	.93	1.00	1.00	1.00	1.00	150
5	.92	.94	.92	.95	.95	.98	.98	1.00	1.00	.90	.95	.96	.91	.98	.99	.99	1.00	1.00	1.00	1.00	250
b_j: minimum no. of weapons to be assigned to target j			30							40				50	70	35				10	
u_j: military value of target j	60	50	50	75	40	100	35	30	25	150	30	45	125	200	200	130	100	100	100	150	

Table 2.2 Weapon Assignments (after Rounding to Nearest Integer)

										x_{ij}: Weapons of Type i Assigned to Target j											
										Targets											Total No. of
Weapon	1	2	3	4	5	6	7	8	9	10	11	12	13	14	15	16	17	18	19	20	Weapons Assigned
1		16				100	38	26	20												200
2				23	20										25	31	2				101
3															45		76	56	62	61	300
4	50	46	47													4					150
5										50	57	39	50	57							250
Total weapons assigned to target	50	62	47	23	20	100	38	26	20	50	57	39	50	57	70	35	78	56	62	61	

A solution of the nonlinear programming model is given in Table 2.2. Values determined by the algorithm and shown in this table have all been rounded to the nearest integer. Note that the assignment of 101 weapons of type 2 exceeds the availability. In this case we check the solution values to find which one has been rounded the most and find that $x_{2,17} = 2$ was rounded from 1.627, so we set $x_{2,17} = 1$. A rule for rounding in this problem could be to round to the nearest integer and then check all the constraints to see if they are satisfied. If they are not, perform manual adjustments. There can be no simple rule, since there are both upper and lower limits on assignments.

2.4 SOURCE OF PROBLEM AND REFERENCE

Source

Work on this type of weapons assignment problem has been performed at the Research Analysis Corporation for several years. Work by W. Eckhart, M. Brush, and R. Gramann has resulted in a linear programming model for weapons assignment that uses linear approximations to nonlinear functions such as those given in this chapter. The two models described in this chapter were formulated by W. Charles Mylander, III, who discussed one of them in [1].

Reference

[1] W. C. Mylander, III, "Applied Mathematical Programming," *Proc. U.S. Army Operations Res. Symp.*, Part 1 (March 1965).

3
BID EVALUATION

3.1 DESCRIPTION OF BID EVALUATION PROBLEM

There are many variations of the bid evaluation problem. The following description is general enough to include most of the elements that make it difficult to solve.

A company wishes to purchase a specified number of units of an item. It obtains bids from n vendors, each of whom cannot necessarily supply the total amount. The vendors submit bids indicating their prices as functions of the amounts purchased. Such bids usually reflect setup costs and decreasing unit costs (depending on order sizes) as well as maximum and minimum quantities. The problem is to choose the amount to be purchased from each vendor so as to minimize the total cost for purchasing the required items.

For specificity, consider the following bid evaluation problem. A buyer wishes to purchase 239,600,480 units of a specific commodity. Five vendors, A through E, have submitted bids to supply certain quantities, but no one vendor can supply the total desired by the buyer. In their proposals all of

Vendor	Setup Cost	Unit Price	Quantity (Units)
A	$3,855.84	$0.061150	0–33,000,000
B	125,804.84	0.068099	22,000,000–70,000,000
		0.066049	70,000,001–100,000,000
		0.064099	100,000,001–150,000,000
		0.062119	150,000,001–160,000,000
C	13,456.00	0.06219	0–165,600,000
D	6,583.98	0.072488	0–12,000,000
E	0	0.070150	0–42,000,000
		0.068150	42,000,001–77,000,000

Description of Bid Evaluation Problem

the bidders except one have included setup costs, which are independent of quantity sold. The buyer is also faced with the problem of scaled costs; that is, a certain unit cost is quoted for a certain minimum quantity to be sold, a lower unit cost is quoted for a larger quantity to be sold, and so on. The relevant information is included in the table at the foot of page 28.

Let $\varphi_1(x_1)$ denote the cost of x_1 *thousands of units* supplied by vendor A, $\varphi_2(x_2)$ denote the cost of x_2 thousands of units supplied by vendor B, etc. The total cost is $\varphi(x) = \varphi_1(x_1) + \varphi_2(x_2) + \varphi_3(x_3) + \varphi_4(x_4) + \varphi_5(x_5)$. The following transforms the information in the previous table into cost functions. The "level" information on the right will be used later.

			Level
$\varphi_1(x_1) =$	$\begin{cases} 0 \\ 3{,}855.84 + 61.150 x_1 \end{cases}$	$\begin{array}{l} x_1 = 0 \\ 0 < x_1 \leq 33{,}000 \end{array}$	1 2
$\varphi_2(x_2) =$	$\begin{cases} 0 \\ 1{,}623{,}982.84 \\ 125{,}804.84 + 68.099 x_2 \\ 269{,}304.84 + 66.049 x_2 \\ 464{,}304.84 + 64.099 x_2 \\ 761{,}304.84 + 62.119 x_2 \end{cases}$	$\begin{array}{l} x_2 = 0 \\ 0 < x_2 \leq 22{,}000 \\ 22{,}000 < x_2 \leq 70{,}000 \\ 70{,}000 < x_2 \leq 100{,}000 \\ 100{,}000 < x_2 \leq 150{,}000 \\ 150{,}000 < x_2 \leq 160{,}000 \end{array}$	1 2 3 4 5 6
$\varphi_3(x_3) =$	$\begin{cases} 0 \\ 13{,}456.0 + 62.019 x_3 \end{cases}$	$\begin{array}{l} x_3 = 0 \\ 0 < x_3 \leq 165{,}000 \end{array}$	1 2
$\varphi_4(x_4) =$	$\begin{cases} 0 \\ 6{,}583.98 + 72.488 x_4 \end{cases}$	$\begin{array}{l} x_4 = 0 \\ 0 < x_4 \leq 12{,}000 \end{array}$	1 2
$\varphi_5(x_5) =$	$\begin{cases} 70.150 x_5 \\ 84{,}000.00 + 68.150 x_5 \end{cases}$	$\begin{array}{l} 0 \leq x_5 \leq 42{,}000 \\ 42{,}000 < x_5 \leq 77{,}000 \end{array}$	1 2

Note that zero units may be purchased from vendor B; otherwise no positive number of units less than 22,000,000 may be purchased. We incorporate this condition in $\varphi_2(x_2)$ by requiring $\varphi_2(x_2) = \$1{,}623{,}982.84$ for $0 < x_2 \leq 22{,}000$, noting that this is just the sum of the setup cost plus the cost of 22,000,000 units at 0.068099 per unit.

The bid evaluation problem may now be written as the following multi-level, fixed-charge problem.

Choose $x_1, \ldots, x_5$ to

$$\text{minimize } \varphi_1(x_1) + \varphi_2(x_2) + \varphi_3(x_3) + \varphi_4(x_4) + \varphi_5(x_5)$$

subject to

$$x_1 + x_2 + x_3 + x_4 + x_5 = 239{,}600.48,$$
$$0 \leq x_1 \leq 33{,}000$$
$$0 \leq x_2 \leq 160{,}000$$
$$0 \leq x_3 \leq 165{,}000$$
$$0 \leq x_4 \leq 12{,}000$$
$$0 \leq x_5 \leq 42{,}000.$$

3.2 BRANCH AND BOUND METHOD FOR SOLUTION

This problem, although the objective function and constraints have only linear elements, qualifies as a nonlinear programming problem because the total objective function varies nonlinearly as a function of the number of items purchased. It is not, however, a nonlinear programming problem as defined in Chapter 1, because of two mathematical characteristics.

First, since setup costs (usually referred to as "fixed charges") are allowed, the objective function is discontinuous. That is, if no item is purchased from a particular vendor, there is no cost. If one is purchased, there is a jump in the cost function of the amount of the fixed charge. Second, there are abrupt rather than continuous changes in unit cost per item as a function of the total number purchased. Although the total cost function is continuous, it is nonsmooth, or, in mathematical terms, its derivatives are discontinuous at the points where the unit costs change.

As stated, the bid evaluation problem is an example of a multilevel, fixed-charge problem consisting of a cost function with piecewise linear segments. Algorithms for solving standard linear programming problems are not applicable to the problem as it is stated. Although the objective function is not strictly linear, nonlinear programming algorithms do not apply either.

Because of its piecewise smooth structure, this problem can be expanded into a number of linear programming subproblems by restricting each of the variables to one of the intervals on which its objective function is linear. By solving all of the possible subproblems the solution of the large problem may be found. There are two levels at which purchases can be made from vendor A, six for vendor B, two for vendor C, two for vendor D, and two for vendor E. The number of possible programming subproblems is $2 \times 6 \times 2 \times 2 \times 2 = 96$. Although for problems of this size there is no difficulty in solving this number of subproblems on a computer, the combinatorial possibilities inherent in bid evaluation problems are usually much too large to allow their solution. In recent years, research on the "traveling salesman" problem has stimulated the development of a technique for reducing the number of subproblems necessary to solve in order to find the optimum of the original problem. The name given to this technique is "branch and bound."

Branch and bound techniques are based on the capability of partitioning sets of possible subproblems into subsets and associating a lower bound with each subset that applies to every subproblem in that subset. If any lower bound exceeds a known feasible solution value, then that entire subset of possibilities can be ignored. The remaining subsets are then partitioned into smaller subsets. Note that the lower bounds generated for the new

Branch and Bound Method for Solution

partitioning must be higher (or equal) to that of the subset from which they were created. The process terminates when one of the subproblems yields a solution value less than or equal to the current lower bound for every other subset.

For different problems the implementation, the generation of the lower bounds, and the method of partitioning all vary. For multilevel, fixed-charge problems, a class of problems of which the bid evaluation problem is just one example, a branch and bound algorithm was developed in [2] and is summarized here.

To explain the branch and bound technique used to solve the bid evaluation, we introduce the following notation. The ith subset of subproblems of the bid evaluation problem is described by a five-component vector, $s_i = (j_1^i, \ldots, j_5^i)$. Each component of s_i can take on as many values as there are levels of purchasing available from the corresponding vendor, plus 1. A 0 in the kth component indicates that the subset contains no restrictions on the level at which the kth vendor can be purchased. A 1 in the kth component means that the kth vendor is to be purchased at his first-interval level. A 2 in the kth component means that vendor k is to be purchased at his second-interval level, etc. Thus the vector $(0, 5, 1, 2, 0)$ indicates the subproblems in which vendor 1 can have any level of purchase, vendor 2 can be purchased between 100,000 and 150,000 items, etc. Obviously the subset indicated by $(0, 0, 0, 0, 0)$ is the set of all possible subproblems. The possibilities or levels in each component have been indicated in the rightmost column of the definition of the programming problem.

The bound for each subset is computed in the following way. A programming problem is associated with each subset. The variables with nonzero components in the s_i vector are restricted to their indicated level, and their costs are the corresponding linear functions. The cost for variables whose s_i components are zero is the linear function with largest slope that underestimates the piecewise linear cost associated with that variable in its total interval, and that passes through the origin. Or, mathematically, let $c_i = \max \{c \mid cx_i \leq \varphi_i(x_i) \text{ for } 0 \leq x_i \leq d_i\}$, where d_i is the maximum amount that could be purchased from that vendor (the upper limit on x_i). Then $c_i x_i$ is the cost term of that variable in the subproblem associated with s_i. Thus the solution to the programming problem generated in this manner is lower than the solution to any subproblem in that subset. This is the lower bound associated with that subset.

To illustrate this we exhibit the programming problem associated with $s = (0, 5, 1, 2, 0)$.

Choose values of $x_1, \ldots, x_5$ to

minimize $61.267x_1 + 464{,}304.84 + 64.099x_2 + 0x_3$
$\qquad\qquad + 6{,}583.98 + 72.488x_4 + 69.241x_5$

subject to
$$x_1 + x_2 + x_3 + x_4 + x_5 = 239{,}600.48,$$
$$0 \le x_1 \le 33{,}000$$
$$100{,}000 \le x_2 \le 150{,}000$$
$$x_3 = 0$$
$$0 \le x_4 \le 12{,}000$$
$$0 \le x_5 \le 77{,}000.$$

Because the marginal costs in this problem are nonincreasing, the linear function with maximum slope that passes through the origin and underestimates any piecewise linear cost function is easily computed by dividing by the maximum number the total cost for any vendor at the maximum number of items purchasable and using the term as the coefficient for the variable in the objective function. Thus, since x_1 has a zero coefficient in the s vector, its objective function coefficient is

$$c_1 = \frac{3{,}855.84 + 61.150 \times 33{,}000}{33{,}000} \doteq 61.267.$$

Similarly,
$$c_5 = \frac{84{,}000.00 + 68.150 \times 77{,}000}{77{,}000} \doteq 69.241.$$

Two numbers are obtained from the solution to the programming problem associated with s. The first, its solution value, we denote by $P(s)$. This is the lower bound associated with that subset. Let $x(s)$ denote the solution vector for that problem and $x_i(s)$ denote its ith component. Since it is a feasible vector, its cost value using the true cost function $\varphi(s)$ is an upper bound for the solution value to the original programming problem. Thus, if any subset has a lower bound on a subset higher than any upper bound on the original problem given by any feasible point, that subset can be discarded since it cannot contain any subproblem better than those in other subsets. This simple fact allows the exclusion of many of the combinatorial possibilities and makes the algorithm efficient.

The next requirement for the branch and bound technique is the rule for deciding which of the existing subsets to partition further in the search for the optimum subproblem. For the bid evaluation problem the rule is to partition the subset with the lowest lower bound.

The final rule is how to partition that subset. The rule is to choose one of the components that is zero and expand the subset into all the possibilities with respect to that variable. Thus one possible partitioning of (0, 5, 1, 2, 0)

Solution of Example Problem

is (1, 5, 1, 2, 0) and (2, 5, 1, 2, 0). The only other possibility is to partition with respect to the fifth component. The choice of which component is more of an art than a science. The (effectively) arbitrary fixing of the order in which the variables are considered is clearly not as efficient as an order determined by the particular problem being solved. In fact, no rigid order is desirable; rather, the choice of the next variable to be considered should be made only when such a variable is about to be chosen. An effective rule used for choosing the variable x_i on which s is to be partitioned is as follows. For the solution vector $x(s)$ yielding the solution value $P(s)$ we choose x_i such that

$$\varphi_i[x_i(s)] - c_i x_i(s) = \max_l \{\varphi_l[x_l(s)] - c_l x_l(s)\}.$$

In this case the index $l = 1, \ldots, 5$. That is, x_i is the variable for which $c_i x_i$ is the worst approximation to $\varphi_i(x_i)$ at the solution to the problem yielding solution value $P(s)$. This procedure attempts to increase the lower bounds as quickly as possible so that fewer subsets need be examined and partitioned.

Assume that the indicated partitioning is made. The next step is to establish a lower bound for (1, 5, 1, 2, 0). This is done by the method indicated above. Several efficiencies are available at this point. If the lower bound for the subset is higher than the current lowest upper bound for the solution of the original problem, then that subset may be removed from further consideration, as stated above. Next, a lower bound is established for (2, 5, 1, 2, 0). Often a full iteration is not required to discard a subset of possibilities. This can occur if the method used to solve the subproblems yields the information that the optimum is higher than the current lowest upper bound before the subproblem is completely solved. Using a dual feasible method (SUMT) this efficiency was incorporated in the algorithm. Other tricks for discarding subsets of possibilities may be used at this point in a general branch and bound algorithm.

3.3 SOLUTION OF EXAMPLE PROBLEM

Having outlined the technique used for the bid evaluation problem, we now solve the example problem given at the start of this section. Figure 3.1 shows graphically the steps described.

ITERATION 1

$$s_1 = (0, 0, 0, 0, 0).$$

For this first problem every variable is allowed to vary from zero to its upper bound. The objective function costs are the linear underestimates for

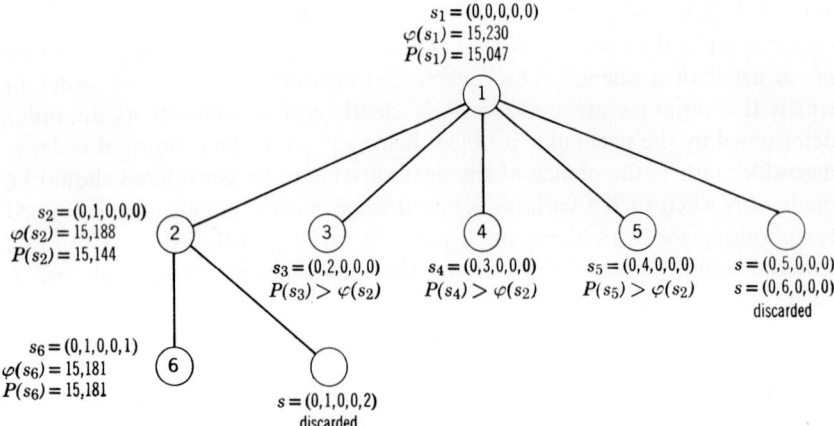

Figure 3.1 Branch and Bound Diagram for Example Problem

every variable. The solution for this problem is

$$x_1 = 33{,}000$$
$$x_2 = 41{,}000.48$$
$$x_3 = 165{,}600$$
$$x_4 = 0$$
$$x_5 = 0.$$

The lower bound for all subproblems is \$15,047,672.88. An upper bound on the final optimum is \$15,230,155.03, since for the x of the first iteration $\varphi(x)$ yields this value.

ITERATION 2

Using the rule discussed previously, the variable x_2 is chosen to define the partitioning

$$s_2 = (0, 1, 0, 0, 0).$$

The solution to the corresponding problem is

$$x_1 = 33{,}000.00$$
$$x_2 = 0$$
$$x_3 = 165{,}600.00$$
$$x_4 = 0$$
$$x_5 = 41{,}000.48.$$

Solution of Example Problem

The solution to this problem is $15,144,569.40, a higher lower bound to the optimum, and the value using the real costs at the above point is $15,188,421.30, an upper bound.

ITERATION 3

$$s_3 = (0, 2, 0, 0, 0).$$

The lower bound for the solution of this problem was higher than the feasible upper bound generated at Iteration 2. Hence all subproblems emanating from this were ignored.

ITERATION 4

$$s_4 = (0, 3, 0, 0, 0).$$

Again, a higher lower bound than the upper bound generated by Iteration 2.

ITERATION 5

$$s_5 = (0, 4, 0, 0, 0).$$

Again, a higher lower bound was obtained than the upper bound generated by Iteration 2. Using the dual bounds available from SUMT the possibilities given by (0, 5, 0, 0, 0) and (0, 6, 0, 0, 0) were discarded without requiring a complete iteration.

ITERATION 6

We now must consider partitions of the subset indicated by (0, 1, 0, 0, 0) since Iterations 3–5 have eliminated the possibility that vendor 2 will be purchased at any level except zero. Using the selection rule discussed previously, x_5 is chosen as the next variable to be explored. Hence $s_6 = (0, 1, 0, 0, 1)$.
The solution to this is

$$x'_1 = 33{,}000.00$$
$$x'_2 = 0$$
$$x'_3 = 165{,}600.00$$
$$x'_4 = 0$$
$$x'_5 = 41{,}000.48.$$

Again, using the dual bound obtained from SUMT, the possibility (0, 1, 0, 0, 2) was discarded. Hence (0, 1, 0, 0, 1) represents all possible solutions to the original problem. Also, at the solution vector above, the upper and lower bounds are equal to $15,181,791.91. Hence no further partitioning is required and the solution has been obtained.

The effectiveness of the algorithm is illustrated by this solution to the bid evaluation problem. Of the 96 subproblems that could possibly need to be solved, only six complete subproblem solutions were required.

For larger problems effective use of branch and bound techniques may be even more valuable. A description of the general algorithm to handle multilevel, fixed-charge problems is contained in [2]. An explanation of branch and bound techniques in general is contained in [1].

3.4 SOURCE OF PROBLEM AND REFERENCES

Source

A. P. Jones and R. M. Soland worked with the bid evaluation model in connection with their branch and bound research presented in [2].

References

[1] N. Agin, "Optimum Seeking with Branch and Bound," *Management Sci.*, **13**, B176–B185 (1966).

[2] A. P. Jones and R. M. Soland, "A Branch and Bound Algorithm for Multi-level Fixed Charge Problems," RAC-TP-285 Research Analysis Corporation, McLean, Va.

4
ALKYLATION PROCESS OPTIMIZATION

In this chapter we describe a model for optimization of the operation of a chemical process common in the petroleum industry. The model seeks to determine the optimum set of operating conditions for the process, based on a mathematical model of the process, a profit function to be maximized, and a set of starting conditions. Most chemical processes can be represented by nonlinear relationships without discontinuities, and they are usually constrained by numerous restrictions on the operating ranges of the variables. The interrelationships among the variables are sufficiently complicated so that changing one variable usually results in changes in a number of the other variables.

The model was described by Sauer, Colville, and Burwick [3], and the process relationships used in the model are based on those given by Payne [2]. The solution procedure used by Sauer, Colville, and Burwick for optimizing the model is a reduction of the nonlinear problem to a series of linear programming problems, which is described by Colville [1]. We have formulated the model as a direct nonlinear programming model with mixed nonlinear inequality and equality constraints and a nonlinear criterion function. The formulation is described in this chapter.

4.1 DESCRIPTION OF ALKYLATION PROCESS MODEL

Description of the Process and Variables

A simplified process flow diagram of an alkylation process is given in Figure 4.1. There is a reactor in which olefin feed and isobutane makeup are introduced. Fresh acid is added to catalyze the reaction, and spent acid is withdrawn. The hydrocarbon product from the reactor is fed into a

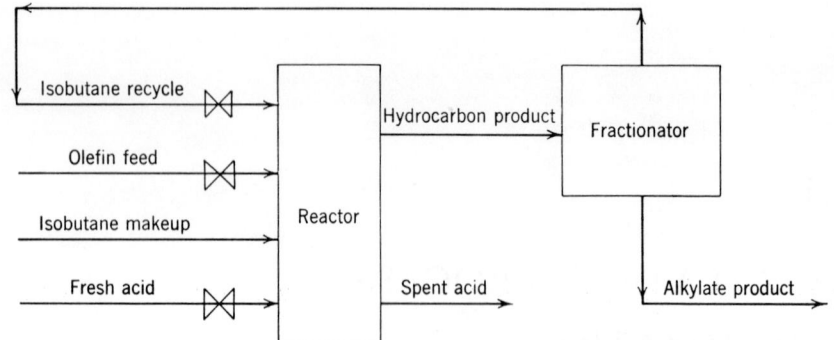

Figure 4.1 Simplified Alkylation Process Flow Diagram

fractionator, and isobutane is taken from the top of the fractionator and recycled back into the reactor. Alkylate product is withdrawn from the bottom of the fractionator. Several of the simplifying assumptions are that the olefin feed is 100 per cent butylene, isobutane makeup and isobutane recycle are 100 per cent isobutane, and fresh acid strength is 98 per cent by weight.

Payne [2] discusses the process variables and their relationships with each other. Some of the relationships involve material balances, while some are correlations between variables within certain ranges, described by linear or nonlinear regressions. We shall develop equality constraints for material balances, and inequality constraints for regression relationships.

It is convenient to define independent and dependent variables in formulating the model, although mathematically the nonlinear programming problem treats the variables alike. The independent variables are the controllable or "knob" variables, the values of which are determined by the operator by changing set points on automatic control instruments. On the process flow diagram these variables are indicated by butterfly valves (-⋈-). Changes in the values of these independent variables induce changes throughout the process. The independent variables are the olefin feed rate in barrels per day, the isobutane recycle in barrels per day, and the fresh acid addition rate in thousands of pounds per day. There are other independent variables, not in the model, which we assume have been appropriately taken care of, such as relative humidity of outside air and temperature of cooling water in the process.

The dependent variables can be divided into three classes: (a) economically significant variables, (b) performance indices, and (c) supporting variables, defined and used when building the model. The economically significant dependent variables are alkylate yield in barrels per day and isobutane makeup in barrels per day. The other dependent variables are acid strength

Description of Alkylation Process Model

by weight per cent, motor octane number (also economically significant), external isobutane-to-olefin ratio, acid dilution factor, and F-4 performance number.

Relationships Used in Determining Constraints

We start off by defining the 10 variables to be considered in the model, which have already been mentioned and are mathematically related in this section. We define

x_1 = olefin feed (barrels per day),
x_2 = isobutane recycle (barrels per day),
x_3 = acid addition rate (thousands of pounds per day),
x_4 = alkylate yield (barrels per day),
x_5 = isobutane makeup (barrels per day),
x_6 = acid strength (weight per cent),
x_7 = motor octane number,
x_8 = external isobutane-to-olefin ratio,
x_9 = acid dilution factor,
x_{10} = F-4 performance number.

Values to be taken on by the variables are all bounded from below and above. The independent variables x_1, x_2, and x_3 and the dependent variables x_4 and x_5 have limitations imposed by the economic situation under analysis. For example, only 2000 barrels per day of olefin feed, x_1, may be available for use in the process. These bounds will be included as constraints in the model. Similarly, the performance indices are required to lie within certain specified ranges because of the physical relationships of the process, and these bounds will be included as constraints.

We give the equations for the dependent variables as functions of independent variables and of other dependent variables. The alkylate yield, x_4, is a function of the olefin feed, x_1, and the external isobutane-to-olefin ratio, x_8. The relationship is determined by a nonlinear regression holding at reactor temperatures between 80 to 90°F and reactor acid strength by weight per cent of 85 to 93. The regression equation is

$$x_4 = x_1(1.12 + .13167x_8 - .00667x_8^2).$$

The isobutane makeup, x_5, can be determined by a volumetric reactor balance. The alkylate yield, x_4, equals the olefin feed, x_1, plus the isobutane makeup, x_5, less shrinkage. The volumetric shrinkage can be expressed as .22 volume per volume of alkylate yield. The balance is then

$$x_4 = x_1 + x_5 - .22x_4,$$

or

$$x_5 = 1.22x_4 - x_1.$$

The acid strength by weight per cent, x_6, can be derived from an equation that expresses acid addition rate, x_3, as a function of alkylate yield, x_4, acid dilution factor, x_9, and acid strength by weight per cent, x_6. The addition acid is assumed to have acid strength of 98. The equation is

$$1000x_3 = \frac{(x_4)(x_9)(x_6)}{(98 - x_6)}.$$

Rearranging, we obtain acid strength as a function of acid addition rate, alkylate yield, and acid dilution factor:

$$x_6 = \frac{98{,}000x_3}{x_4 x_9 + 1000x_3}.$$

The motor octane number, x_7, is a function of the external isobutane-to-olefin ratio, x_8, and the acid strength by weight per cent, x_6. The relationship holds for the same reactor temperatures and acid strengths as for alkylate yield, x_4. The equation determined by nonlinear regression is

$$x_7 = 86.35 + 1.098x_8 - .038x_8^2 + .325(x_6 - 89).$$

The external isobutane-to-olefin ratio, x_8, is equal to the sum of the isobutane recycle, x_2, and the isobutane makeup, x_5, divided by the olefin feed, x_1. The equation is

$$x_8 = \frac{x_2 + x_5}{x_1}.$$

The acid dilution factor, x_9, can be expressed as a linear function of the F-4 performance number, x_{10}. A curve is approximated by the linear regression equation

$$x_9 = 35.82 - .222x_{10}.$$

The last dependent variable is the F-4 performance number, x_{10}, which may be expressed as a linear function of the motor octane number, x_7. The linear regression equation is

$$x_{10} = -133 + x_7.$$

The above relationships give the dependent variables in terms of the independent variables and the other dependent variables. All of the relationships must hold for the process to be in balance. In addition to the above

Table 4.1 Lower and Upper Bounds on Selected Dependent Variables

Dependent Variable	Minimum Limit	Maximum Limit
x_6, acid strength (weight per cent)	85	93
x_7, motor octane number	90	95
x_8, external isobutane-to-olefin ratio	3	12
x_9, acid dilution factor	1.2	4
x_{10}, F-4 performance number	145	162

Description of Alkylation Process Model

relationships, there are lower and upper bounds to be imposed on the variables. The independent variables have these bounds imposed by the capability of the physical plant and/or the economic situation being analyzed. These will be specified in the example. The dependent variables alkylate yield, x_4, and isobutane makeup, x_5, also are affected by the economic situation and other general conditions. But the dependent variables x_6, x_7, x_8, x_9, and x_{10} have bounds that are directly related to the physical process. Table 4.1 shows the minimum and maximum limits for these variables.

Profit Function

The profit function is defined in terms of alkylate product or output value minus feed and recycle costs. Operating costs not reflected in the function we assumed not to vary among possible process setups.

Define the value and cost parameters to be used in the profit function:

c_1 = alkylate product value (dollars per octane-barrel),
c_2 = olefin feed cost (dollars per barrel),
c_3 = isobutane recycle costs (dollars per barrel),
c_4 = acid addition cost (dollars per thousand pounds),
c_5 = isobutane makeup cost (dollars per barrel).

The total profit per day, to be maximized, is

$$\text{Profit} = c_1 x_4 x_7 - c_2 x_1 - c_3 x_2 - c_4 x_3 - c_5 x_5.$$

Specification of Model

Define lower and upper bounds on the variables

$x_j^{(l)}$ = lower bound on the jth variable,
$x_j^{(u)}$ = upper bound on the jth variable,

where $j = 1, \ldots, 10$.

Regression analysis was used to formulate the relationships for x_4, x_7, x_9, and x_{10} in terms of the other variables. Exact models were used for the relationships for x_5, x_6, and x_8. For the former variables we use two inequality constraints that specify a range within which the true value can be approximated by the estimated value. For the latter variables one equality constraint is used.

Thus for the relationship

$$x_4 = f(x_1, x_8)$$

we use

$$f(x_1, x_8) - d_{4_l} x_4 \geq 0,$$
$$-f(x_1, x_8) + d_{4_u} x_4 \geq 0,$$

where d_{4_l} and d_{4_u} are the lower and upper values establishing the percentage difference of the estimated from the true value. An example that illustrates how these inequalities work may be seen by setting $d_{4_l} = \frac{9}{10}$ and $d_{4_u} = \frac{10}{9}$. The inequalities

$$\tfrac{9}{10}x_4 \le f(x_1, x_8) \le \tfrac{10}{9}x_4$$

reduce to

$$f(x_1, x_8) - \tfrac{9}{10}x_4 \ge 0,$$
$$-f(x_1, x_8) + \tfrac{10}{9}x_4 \ge 0.$$

With these preliminaries taken care of, we write the nonlinear programming model for maximizing profit per day of the alkylation process by setting the independent variables equal to the optimal values as follows. Choose x_j ($j = 1, \ldots, 10$) to

$$\text{maximize } c_1 x_4 x_7 - c_2 x_1 - c_3 x_2 - c_4 x_3 - c_5 x_5$$

subject to the constraints

$$x_j^{(l)} \le x_j \le x_j^{(u)}, \quad j = 1, \ldots, 10,$$
$$[x_1(1.12 + .13167 x_8 - .00667 x_8^2)] - d_{4_l} x_4 \ge 0,$$
$$-[x_1(1.12 + .13167 x_8 - .00667 x_8^2)] + d_{4_u} x_4 \ge 0,$$
$$[86.35 + 1.098 x_8 - .038 x_8^2 + .325(x_6 - 89)] - d_{7_l} x_7 \ge 0,$$
$$-[86.35 + 1.098 x_8 - .038 x_8^2 + .325(x_6 - 89)] + d_{7_u} x_7 \ge 0,$$
$$[35.82 - .222 x_{10}] - d_{9_l} x_9 \ge 0,$$
$$-[35.82 - .222 x_{10}] + d_{9_u} x_9 \ge 0,$$
$$[-133 + 3x_7] - d_{10_l} x_{10} \ge 0,$$
$$-[-133 + 3x_7] + d_{10_u} x_{10} \ge 0,$$
$$1.22 x_4 - x_1 - x_5 = 0,$$
$$\frac{98{,}000 x_3}{x_3 x_9 + 1{,}000 x_3} - x_6 = 0,$$
$$\frac{x_2 + x_5}{x_1} - x_8 = 0.$$

The final element to be mentioned is the starting values that are input to the model. These represent a balanced or nearly balanced process that engineers have developed, which should be a feasible solution satisfying the constraints. It is not absolutely necessary, for some nonlinear programming procedures can determine their own feasible solutions, but good starting values can be very helpful in solving the nonlinear programming problem.

4.2 EXAMPLE OF ALKYLATION PROCESS MODEL APPLICATION

In this section we given the necessary data for an example of the model just described and discuss solution of the problem. The example is taken

Example of Alkylation Process Model Application

Table 4.2 Lower and Upper Bounds on Variables, and Starting Values

Variable	Lower Bound	Upper Bound	Starting Value
x_1, olefin feed (barrels per day)	0	2,000	1,745
x_2, isobutane recycle (barrels per day)	0	16,000	12,000
x_3, acid addition rate (thousands of pounds per day)	0	120	110
x_4, alkylate yield (barrels per day)	0	5,000	3,048
x_5, isobutane makeup (barrels per day)	0	2,000	1,974
x_6, acid strength (weight per cent)	85	93	89.2
x_7, motor octane number	90	95	92.8
x_8, external isobutane-to-olefin ratio	3	12	8
x_9, acid dilution factor	1.2	4	3.6
x_{10}, F-4 performance number	145	162	145

from Sauer, Colville, and Burwick [3]. Lower and upper bounds on the variables are given in Table 4.2, which includes the bounds to be used in the particular situation being studied in addition to those previously specified for the physical process. Also given in Table 4.2 are starting values for the optimization procedure.

Parameters for profit from the sale of alkylate and costs of inputs required for production are given in Table 4.3. Using the starting values from Table 4.3,

$$\text{Profit} = (\$.063)(3,048)(92.8) - (\$5.04)(1,745) - (\$.035)(12,000)$$
$$- (\$10.00)(110) - (\$3.36)(1,974)$$
$$= \$872.$$

The final input parameters to be specified are the permissible error relationships for the inequality constraints on the regression relationships. Table 4.4 gives the lower and upper deviation parameters.

Table 4.3 Values of Profit and Cost Parameters

Profit and Cost Parameter	Value
c_1, alkylate product value	$.063 per octane-barrel
c_2, olefin feed cost	$5.04 per barrel
c_3, isobutane recycle cost	$.035 per barrel
c_4, acid addition cost	$10.00 per thousand pounds
c_5, isobutane makeup cost	$3.36 per barrel

Table 4.4 Values of Deviation Parameters

Deviation Parameter	Value
d_{4_l}	99/100
d_{4_u}	100/99
d_{7_l}	99/100
d_{7_u}	100/99
d_{9_l}	9/10
d_{9_u}	10/9
d_{10_l}	99/100
d_{10_u}	100/99

The data in Tables 4.2, 4.3, and 4.4 are sufficient to allow application of the model described in the previous section. Values of the independent and dependent variables that maximize profit subject to the constraints are given in Table 4.5. Also listed are lower and upper bounds and starting values. The profit associated with the optimal solution is $1769 per day, an increase of $897 over that of the starting value.

Isobutane recycle, x_5, is at the upper limit in the optimal solution given above. To test the sensitivity of profits of the process to an increase in the availability of isobutane makeup, we increase the upper limit of x_5 by 10 per cent to 2200 barrels. We also arbitrarily increase the upper bound on fractionation capacity by 25 per cent to 20,000, to allow for more isobutane recycle if this will balance the process at a higher level of profit. The profit goes to $1946, an increase of $1074 over the starting value. Isobutane recycle x_5 is used at the limiting point of 2200 barrels, and isobutane recycle goes to 17,396 barrels, which shows the necessity for increasing the fractionation capacity to balance the increased isobutane makeup.

Table 4.5 Optimal Solution of Example Problem

Variable	Lower Bound	Optimum Value	Upper Bound	Starting Value
x_1	0	1,698	2,000	1,745
x_1	0	15,818	16,000	12,000
x_3	0	54.1	120	110
x_4	0	3,031	5,000	3,048
x_5	0	2,000	2,000	1,974
x_6	85	90.1	93	89.2
x_7	90	95.0	95	92.8
x_8	3	10.5	12	8
x_9	1.2	1.6	4	3.6
x_{10}	145	154	162	145

References

[1] A. R. Colville, Jr., "Process Optimization Program for Non-linear Programming," unpublished paper, IBM Corporation, December 1964.
[2] R. E. Payne, "Alkylation—What You Should Know about This Process," *Petrol. Refiner*, **37,** 316–329 (1958).
[3] R. N. Sauer, A. R. Colville, Jr., and C. W. Burwick, "Computer Points the Way to More Profits," *Hydrocarbon Process. Petrol. Refiner*, **43,** 84–92 (1964).

5
CHEMICAL EQUILIBRIUM

The problem of determining the chemical composition of a complex mixture under chemical equilibrium conditions has long been of interest. Such problems arise in the analysis of the performance of fuels and propellants and in the synthesis of complex organic compounds.

A mixture of chemical species held at a constant temperature and pressure reaches its chemical equilibrium state concurrently with reduction of the free energy of the mixture to a minimum. This is a consequence of the second law of thermodynamics. The objective function to be minimized in the chemical equilibrium model is the expression of the free energy of the chemical mixture under study. The value of the free energy of the mixture is minimized subject to the chemical reactions possible between species of the mixture.

White, Johnson, and Dantzig [3] formulated the chemical equilibrium problem as a mathematical programming problem with linear mass balance constraints representing the possible chemical combinations of the chemical species of the mixture, and a nonlinear objective function representing the free energy of the mixture (to be minimized). They investigated steepest descent and piecewise linear programming approaches to formulating the the problem. In a second paper [2] they explored the piecewise linear programming problem further. The problem is discussed briefly by Dantzig [1], who used it to illustrate the method of generalized linear programming.

5.1 CHEMICAL EQUILIBRIUM MODEL

Consider a mixture of m chemical elements. It has been predetermined that the m different types of atoms can combine chemically to produce n compounds, where the monotonic atom is regarded for our purpose as a

Example of Chemical Equilibrium Model Application

possible compound. Define

x_j = the number of moles of compound j present in the mixture at equilibrium,

$\bar{x}$ = the total number of moles in the mixture, where $\bar{x} = \sum_{j=1}^{n} x_j$,

a_{ij} = the number of atoms of element i in a molecule of compound j,

b_i = the number of atomic weights of element i in the mixture.

The mass balance relationships that must hold for the m elements are

$$\sum_{j=1}^{n} a_{ij} x_j = b_i, \quad i = 1, \ldots, m \qquad (5.1)$$

and

$$x_j \geq 0, \quad j = 1, \ldots, n. \qquad (5.2)$$

Determination of the composition of the mixture at equilibrium is equivalent to determination of the values of x_j ($j = 1, \ldots, n$) that satisfy (5.1) and (5.2) and also minimize the total free energy of the mixture. The total free energy of the mixture is given by

$$\sum_{j=1}^{n} x_j \left[c_j + \ln \left(\frac{x_j}{\bar{x}} \right) \right], \qquad (5.3)$$

where

$$c_j = \left(\frac{F^0}{RT} \right)_j + \ln P,$$

where (F^0/RT) is the modal standard (Gibbs) free energy function for the jth compound, which may be found in tables, and P is the total pressure in atmospheres.

Thus the nonlinear programming problem is as follows. Choose x_j ($j = 1, \ldots, n$) to minimize the nonlinear objective function (5.3) subject to linear constraints (5.1) and non-negativity restrictions (5.2).

5.2 EXAMPLE OF CHEMICAL EQUILIBRIUM MODEL APPLICATION

We consider the example problem formulated and solved by White, Johnson, and Dantzig [3]. The data are from [3]. We solve the nonlinear programming problem by the sequential unconstrained minimization technique and obtain similar answers, though a remark is warranted that the value of the objective function that we obtain is smaller (thus better) than that given in [3].

The problem considered is the determination of the equilibrium composition resulting from subjecting the compound $\frac{1}{2}N_2H_4 + \frac{1}{2}O_2$ to a temperature of 3500°K and a pressure of 750 psi. In Table 5.1 we show for each compound j of 10 possible compounds (where the monotonic atoms are termed compounds) the Gibbs free energy function $(F^0/RT)_j$, the computed value of c_j for $P = 750$ psi, and the number of atoms of H, N, and O per molecule. The number of atomic weights of H, N, and O in the mixture are assumed to be $b_1 = 2$, $b_2 = 1$, and $b_3 = 1$.

Formulating the nonlinear programming model, the nonlinear objective function to be minimized is

$$x_1\left[-6.089 + \ln\left(x_1 \bigg/ \sum_{j=1}^{10} x_j\right)\right]$$
$$+ \cdots$$
$$+ x_{10}\left[-22.179 + \ln\left(x_{10} \bigg/ \sum_{j=1}^{10} x_j\right)\right],$$

and the linear constraints of the nonlinear programming problem are as follows:

$$x_1 + 2x_2 + 2x_3 + x_6 + x_{10} = 2,$$
$$x_4 + 2x_5 + x_6 + x_7 = 1,$$
$$x_3 + x_7 + x_8 + 2x_9 + x_{10} = 1,$$
$$x_1 \geq 0, \ldots, x_{10} \geq 0.$$

Solving the above nonlinear programming problem, we obtain the values of x_j ($j = 1, \ldots, 10$), the number of moles of the 10 compounds present in the equilibrium mixture, which are given in Table 5.2. These values

Table 5.1 Data on $\frac{1}{2}N_2H_4 + \frac{1}{2}O_2$ at 3500°K, 750 psi

				a_{ij}		
				$i = 1$	$i = 2$	$i = 3$
j	Compound	$(F^0/RT)_j$	c_j	H	N	O
1	H	−10.021	−6.089	1		
2	H_2	−21.096	−17.164	2		
3	H_2O	−37.986	−34.054	2		1
4	N	−9.846	−5.914		1	
5	N_2	−28.653	−24.721		2	
6	NH	−18.918	−14.986	1	1	
7	NO	−28.032	−24.100		1	1
8	O	−14.640	−10.708			1
9	O_2	−30.594	−26.662			2
10	OH	−26.111	−22.179	1		1

Table 5.2 Composition of $\frac{1}{2}N_2H_4 + \frac{1}{2}O_2$ at 3500°K, 750 psi

j	Compound	x_j
1	H	.0407
2	H_2	.1477
3	H_2O	.7831
4	N	.0014
5	N_2	.4853
6	NH	.0007
7	NO	.0274
8	O	.0180
9	O_2	.0373
10	OH	.0969

agree with those obtained in [3]. The corresponding value of the objective function is -47.76.

5.3 SOURCE OF PROBLEM AND REFERENCES

Source

Robert E. Pugh applied the sequential unconstrained minimization technique to this problem, examining the example problem given in [3].

References

[1] G. B. Dantzig, *Linear Programming and Extensions*, Princeton University Press, Princeton, N.J., 1963.
[2] G. Dantzig, S. Johnson, and W. White, "A Linear Programming Approach to the Chemical Equilibrium Problem," *Management Sci.*, **5**, 38–43 (1958).
[3] W. B. White, S. H. Johnson, and G. B. Dantzig, "Chemical Equilibrium in Complex Mixtures," *J. Chem. Phys.*, **28**, 751–755 (1958).

6
STRUCTURAL OPTIMIZATION

In this chapter we describe a nonlinear programming model for optimization of the design of a vertically corrugated transverse bulkhead of an oil tanker. The model determines the design that minimizes the total weight of the transverse bulkhead subject to constraints on performance characteristics and on certain dimensions. The constraints are both linear and nonlinear, and the objective function is nonlinear. The nonlinear programming problem is not convex.

The model was given by Kavlie, Kowalik, and Moe [3]. They cite reference material in their paper. The solution procedure used by Kavlie, Kowalik, and Moe was the sequential unconstrained minimization technique for nonlinear programming with the incorporation of unconstrained minimization techniques of Davidon [1] and Fletcher and Powell [2]. The present writers solved the model using the SUMT program on the IBM 7040, obtaining only slightly different results. The formulation of the model is described in this chapter in terms of the specific bulkhead analyzed by Kavlie, Kowalik, and Moe.

In this chapter we first generally describe the vertical corrugated bulkhead and the design problem, defining the design variables. Then we discuss the constraints and objective function, defining necessary parameters as we go along. The model is given in detail. Results of solution of the nonlinear programming problem are presented.

6.1 DESCRIPTION OF DESIGN PROBLEM

Vertical transverse bulkheads form the lateral walls of the internal compartments of tankers that hold liquid cargo. Longitudinal bulkheads and other structures form the longitudinal walls of the compartments. Corrugated bulkheads have certain design advantages over plane bulkheads, which make

Description of Design Problem

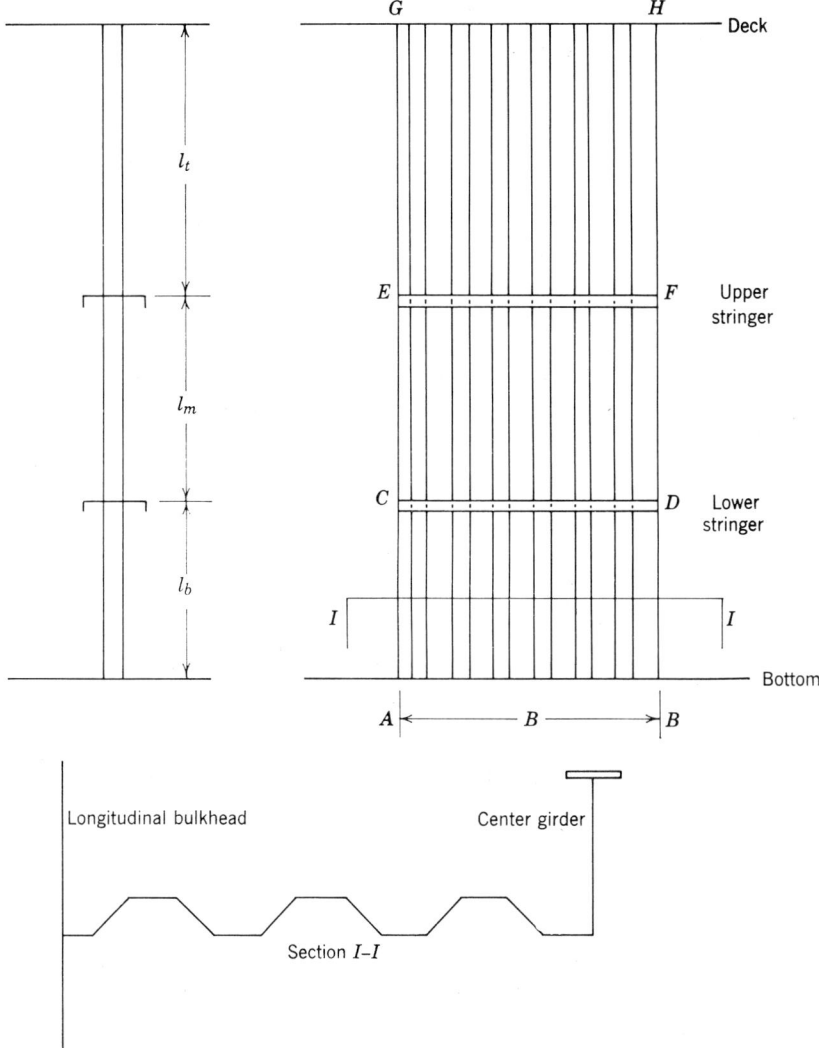

Figure 6.1 Vertical Corrugated Transverse Bulkhead

them candidates for inclusion in tankers. They have been used in some tankers, but are not the usual design.

The corrugated bulkhead to be considered is shown in Figure 6.1. We assumed the shapes of the corrugations to be identical in all of the panels, and the positions of the stringers CD and EF are assumed to be fixed. The lengths of the top, middle, and bottom panels are denoted by l_t, l_m, and l_b,

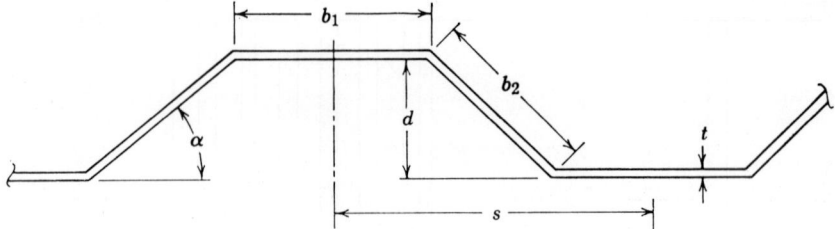

Figure 6.2 Specification of Design Variables

respectively. The width of the panels is denoted by B. It is possible to vary the positions of the stringers, but they are assumed for this problem to be fixed because of the configuration of the neighboring longitudinal bulkheads and other members. An optimization model involving stringer position could include stringers of longitudinal bulkheads.

In Figure 6.2 the basic design variables are shown for one panel. The design variables for all three panels are

b_1 = width of flange (centimeters),
b_2 = length of web (centimeters),
d = depth of corrugation (centimeters),
t_t = thickness of plate in top panel $EFGH$ (centimeters),
t_m = thickness of plate in middle panel $CDEF$ (centimeters),
t_b = thickness of plate in bottom panel $ABCD$ (centimeters).

The corrugations are assumed to be identically shaped in all three panels, but the thicknesses are allowed to vary among panels, thus giving variables t_t, t_m, and t_b.

6.2 DESCRIPTION OF OBJECTIVE FUNCTION

The objective function (to be minimized) is the total weight of the three panels of the corrugated bulkhead. Figures 6.1 and 6.2 show the dimensions to be incorporated in the weight function. We also define

n = number of corrugations,
Γ = weight per unit volume of the material (tons per cubic centimeter).

The total weight of the bulkhead in tons is

$$W = \Gamma n (b_1 + b_2)(t_t l_t + t_m l_m + t_b l_b).$$

The number of corrugations can be expressed in terms of the width of the panel, B, and the width per corrugation, s, by

$$n = \frac{B}{s}.$$

Description of Constraints

Thus the objective function becomes

$$W = \Gamma B(b_1 + b_2)\frac{t_t l_t + t_m l_m + t_b l_b}{s}. \tag{6.1}$$

6.3 DESCRIPTION OF CONSTRAINTS

Figure 6.3 shows the lengths in centimeters of the stiffener spans (panels), l_t, l_m, and l_b; the heights of pressure at the middle of the spans, h_t, h_m, and h_b; and the heights of pressure at the lower ends of the spans, h_{1t}, h_{1m}, and h_{1b}.

Section Modulus

The first three constraints are on the section modulus of each of the three panels. The section modulus is derived by Kavlie, Kowalik, and Moe [3] to be

$$Z = \frac{dt}{2}\left(\frac{b_2}{3} + b_1 e\right) \quad (\text{cm}^3), \tag{6.2}$$

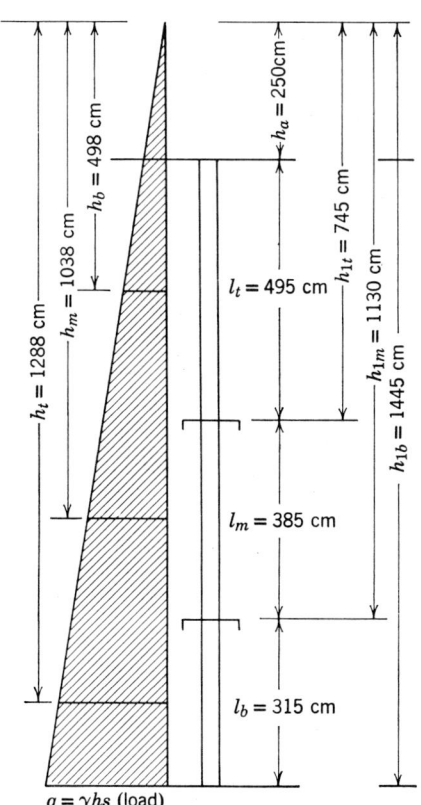

Figure 6.3 Specification of Design Parameters

where e is the effectiveness of the flange (dimensionless). The subscript is dropped from the t because there are three constraints for the three panels.

For each of the three panels, the requirement is that

$$Z \geq Z_{\text{rule}}, \qquad (6.3)$$

where Z_{rule} is derived from specifications given in 1964 rules issued in *Det Norske Veritas* (Section III, Paragraph 6). Each stiffener is assumed to be clamped at the ends with a constant load:

$$q = \gamma h s \quad \text{(kg/cm)},$$

where the specific gravity of fresh water is

$$\gamma = .001 \text{ kg/cm}^3$$

and the geometrical relationship gives

$$s = b_1 + \sqrt{b_2^2 - d^2} \quad \text{(cm)}.$$

In this case the bending moment at the supports is

$$M = \frac{ql^2}{12} = \frac{\gamma h s l^2}{12} \quad \text{(kg-cm)}.$$

The maximum permitted bending stress, $\sigma_b^{\max}$, is 1200 kg/cm². The bending stress is

$$\sigma_b = \frac{M}{Z} \quad \text{(kg/cm}^2\text{)}.$$

Thus, since $\sigma_b \leq \sigma_b^{\max}$,

$$\frac{M}{Z} \leq 1200 \text{ kg/cm}^2$$

$$Z \geq \frac{\gamma h s l^2}{12 \cdot 1200}$$

$$\geq K_1 h s l^2 \quad \text{(cm}^3\text{)}, \qquad (6.4)$$

where $K_1 = \gamma/14{,}400 = .0000000694 \text{ cm}^{-1}$.

Moment of Inertia

The moment of inertia is calculated directly from the section modulus as

$$J = Z \frac{d}{2}. \qquad (6.5)$$

For each of the three panels the requirements given by Kavlie, Kowalik, and Moe [3] are that

$$J \geq 2.2 \sqrt[3]{Z_{\text{rule}}^4}. \qquad (6.6)$$

Plate Thickness

For each of the three panels it is required that

$$t \geq \begin{cases} t^{\min} \\ 3.9b\sqrt{h_1} + K_2, \end{cases} \qquad (6.7)$$

where t = plate thickness (millimeters),
$t^{\min}$ = function of length of ship (centimeters),
$b = \begin{cases} 1.05b_1 \\ 1.05b_2 \end{cases}$ (meters),
h_1 = height of pressure at lower end of panel (meters),
K_2 = corrosion allowance (millimeters).

Geometrical Limitations

The length of the web is constrained to be equal to or greater than the depth of corrugation,

$$b_2 - d > 0, \qquad (6.8)$$

which is obvious from Figure 6.2 if there are to be other than right angles in the corrugations.

6.4 SPECIFICATION OF NONLINEAR PROGRAMMING MODEL

In this section we use the constraint and objective function descriptions given in the two previous sections to formulate a nonlinear programming model. There are 16 constraints, five sets of three each for top, middle, and bottom panels, and one additional constraint.

For all of the parameters of the model we use the notation of the previous sections, but for convenience we define the design variables

$x_1 = b_1$ = width of flange,
$x_2 = b_2$ = length of web,
$x_3 = d$ = depth of corrugation,
$x_4 = t_t$ = thickness of plate in top panel,
$x_5 = t_m$ = thickness of plate in middle panel,
$x_6 = t_b$ = thickness of plate in bottom panel.

Objective Function

From the definition of the objective function (6.1), and using other relationships previously given, the objective function (to be minimized) is

$$W = \Gamma B(x_1 + x_2)(t_t x_4 + t_m x_5 + t_b x_6)[x_1 + (x_2^2 - x_3^2)^{1/2}]^{-1}. \qquad (6.9)$$

Constraints

From (6.2), (6.3), and (6.4), and substituting in the variables $x_1, \ldots, x_6$, we obtain the three constraints on section modulus:

$$\tfrac{1}{6}x_2x_3x_4 + \frac{e}{2} x_1x_3x_4 - K_1h_t l_t^2[x_1 + (x_2^2 - x_3^2)^{1/2}] \geq 0, \quad (6.10)$$

$$\tfrac{1}{6}x_2x_3x_5 + \frac{e}{2} x_1x_3x_5 - K_1h_m l_m^2[x_1 + (x_2^2 - x_3^2)^{1/2}] \geq 0, \quad (6.11)$$

$$\tfrac{1}{6}x_2x_3x_6 + \frac{e}{2} x_1x_3x_6 - K_1h_b l_b^2[x_1 + (x_2^2 - x_3^2)^{1/2}] = 0. \quad (6.12)$$

From (6.2) and (6.5) we obtain the three constraints on moment of inertia:

$$\tfrac{1}{12}x_2x_3^2x_4 + \frac{e}{4} x_1x_3^2x_4 - 2.2(K_1h_t l_t^2)^{4/3}[x_1 + (x_2^2 - x_3^2)^{1/2}]^{1/3} \geq 0, \quad (6.13)$$

$$\tfrac{1}{12}x_2x_3^2x_5 + \frac{e}{4} x_1x_3^2x_5 - 2.2(K_1h_m l_m^2)^{4/3}[x_1 + (x_2^2 - x_3^2)^{1/2}]^{1/3} \geq 0, \quad (6.14)$$

$$\tfrac{1}{12}x_2x_3^2x_6 + \frac{e}{4} x_1x_3^2x_6 - 2.2(K_1h_b l_b^2)^{4/3}[x_1 + (x_2^2 - x_3^2)^{1/2}]^{1/3} \geq 0. \quad (6.15)$$

Noting that the definitions of the terms in (6.7) are not in centimeters, we use (6.7) to obtain the following nine constraints on plate thickness for the three panels:

$$x_4 - t_t^{\min} \geq 0, \quad (6.16)$$

$$10x_4 - [3.9 \cdot 1.05(.01h_{1t})^{1/2}(.01x_1) + 10K_2] \geq 0, \quad (6.17)$$

$$10x_4 - [3.9 \cdot 1.05(.01h_{1t})^{1/2}(.01x_2) + 10K_2] \geq 0, \quad (6.18)$$

$$x_5 - t_m^{\min} \geq 0, \quad (6.19)$$

$$10x_5 - [3.9 \cdot 1.05(.01h_{1m})^{1/2}(.01x_1) + 10K_2] \geq 0, \quad (6.20)$$

$$10x_5 - [3.9 \cdot 1.05(.01h_{1m})^{1/2}(.01x_2) + 10K_2] \geq 0, \quad (6.21)$$

$$x_6 - t_b^{\min} \geq 0, \quad (6.22)$$

$$10x_6 - [3.9 \cdot 1.05(.01h_{1b})^{1/2}(.01x_1) + 10K_2] \geq 0, \quad (6.23)$$

$$10x_6 - [3.9 \cdot 1.05(.01h_{1b})^{1/2}(.01x_2) + 10K_2] \geq 0. \quad (6.24)$$

Finally, from (6.8) we obtain the final constraint on length of web greater than depth of corrugation, which we express as the inequality constraint

$$x_2 - x_3 \geq 0. \quad (6.25)$$

6.5 RESULTS OF SOLUTION

Our solutions do not agree exactly with those of Kavlie, Kowalik, and Moe [3], but they are approximately the same. Like these authors, we solve the problem using two *Det Norske Veritas* design coefficients, 3.9 and 4.25 (for 1964 and 1955, respectively), in the linear plate thickness requirement (6.7), and thus in our constraints (6.17), (6.18), (6.20), (6.21), (6.23), and (6.24). Their solutions and our solutions are given in Table 6.1.

Table 6.1 Optimal Solutions to Structural Optimization Problem

	Starting Point	Kavlie-Kowalik-Moe Optimal Solution [3]		Our Optimal Solution	
		DNV* 1964 (3.9)	DNV 1955 (4.25)	DNV 1964 (3.9)	DNV 1955 (4.25)
b_1	45.8 cm	64.2 cm	55.6 cm	57.8 cm	55.7 cm
b_2	43.2 cm	64.2 cm	63.2 cm	57.8 cm	55.7 cm
d	30.5 cm	36.9 cm	36.7 cm	37.8 cm	37.3 cm
t_t	1.2 cm	1.05 cm	1.05 cm	1.05 cm	1.05 cm
t_m	1.2 cm	1.05 cm	1.05 cm	1.05 cm	1.05 cm
t_b	1.3 cm	1.15 cm	1.15 cm	1.05 cm	1.10 cm
W	6.40 tons	5.29 tons	5.38 tons	5.34 tons	5.44 tons

* Det Norske Veritas.

Kavlie, Kowalik, and Moe obtained a lower total weight than we do for both problems, but our calculations indicate that for both of their optimal solutions the moment of inertia constraint (6.14) is not satisfied. We have found no indication in their paper as to why this would be the case, but perhaps there is an improper coefficient somewhere in the model.

The solutions to the model using the SUMT program required approximately 200 sec on the IBM 7040 to obtain final convergence.

References

[1] W. C. Davidon, "Variable Metric Method for Minimization," Research and Development Report ANL-5990 (Revised), Argonne National Laboratory, U.S. Atomic Energy Commission, 1959.
[2] R. Fletcher and M. J. D. Powell, "A Rapidly Convergent Descent Method for Minimization," *Computer J.*, **7**, 149–154 (1964).
[3] D. Kavlie, J. Kowalik, and J. Moe, "Structural Optimization by Means of Nonlinear Programming," Department of Ship Structures, Technical University of Norway, Trondheim, 1966.

7
LAUNCH VEHICLE DESIGN AND COSTING

7.1 INTRODUCTION

As the nation has progressed farther into the space program, the job of developing space vehicles has been more complex, time-consuming, and costly. Although the technical problems and the meeting of schedules have received first attention in the past, the job of cost prediction and control has become more demanding and is requiring a greater emphasis. As a result of this trend, new and improved tools are required so that today's manager can make quantitative appraisals of the costing problem.

One of the tools receiving a great deal of attention, and the subject of considerable research, is the use of cost models. Cost models typically use the physical characteristics of the subject system as independent variables and include as dependent variables the various costs of the system. Use of these models allows a quantitative approach to many costing problems.

The National Aeronautics and Space Administration has directed and conducted a number of studies in the development of cost models for space vehicles. Most of this effort has been in the determination of launch vehicle cost models, although work is also being done on spacecraft and total program space vehicle cost models. The need for accurate and responsive cost-predictive tools is motivating research in all areas.

Most of the work to date has been to improve the methods of producing cost predictions. The studies have been along the lines of determining cost parameters, developing cost-estimating relationships (CERs), and forming descriptive cost models. It is the objective of this paper to go one step beyond the descriptive cost model and develop an optimizing cost model. In accomplishing this objective we have utilized previous studies as much as possible.

Introduction

Cost-Estimating Relationships

A key element in the development of the subject model is the use of CERs. The CERs relate the design parameters and variables of a launch vehicle to the cost of the launch vehicle or the cost of that particular subsystem of the launch vehicle being considered. The summation of CER terms for each subsystem being considered forms the objective function of the subject model.

The CERs used in the model were developed by a cost study on launch vehicle components conducted by Lockheed Missiles and Space Company [1]. The determination of CERs was accomplished by cost data collection and extrapolation, selection of design parameters, and formulation of equations. The primary data sources were Lockheed in-house cost data files, NASA-supplied Saturn data, previously collected data used in NASA cost models, and data from other contractors. By using this information, and through forecasting and extrapolating, a data base was provided for correlation analysis. The final selection of design parameters and the formulation of cost-estimating equations were performed through correlation analysis. The correlation was performed either by computer, using Lockheed's Weighted Regression Analysis Program (WRAP), or manually for the scarce data cases.

It must be recognized that the CERs were developed from a limited amount of data, and extrapolation for advanced systems is not a precise art. Despite these limitations it is necessary that CERs be formed with reliance on the best data available to enable advanced systems planning.

The need for more accurate CERs is a requirement of all cost models. CERs have been developed by various contractors and agencies, but further work is still required.

Introductory Discussion of Constrained Optimization Model

In order that the model may be seen in as realistic a form as possible, an example launch vehicle is used as the basis. The example vehicle is a three-stage launch vehicle typical of what might be required for an earth-orbiting or earth-escape mission. A detailed description of the example launch vehicle is given in Section 7.2. It is noted, however, that the model is not limited to the example vehicle and that the objective function and constraints can be changed to accommodate any type of vehicle for which CERs can be determined.

Although an example launch vehicle is used, it is our intent to keep the model in as generalized a format as possible and to illustrate the method for developing models for all types of launch vehicles. In general the objective function is a cost function that is to be minimized. The cost function is expressed by CERs as a function of variables that determine the total cost for developing, building, and launching a launch vehicle. The constraints are based on the desired performance of the launch vehicle. A typical kind of over-all constraint is that a launch vehicle must boost a specified payload (PL) to a certain velocity. This is the type of over-all constraint used for the subject model and developed in Section 7.3. To develop all the constraints that express the required physical characteristics of the system, a number of nonlinear and linear relationships are specified.

The model as developed for the example launch vehicle is solved by the sequential unconstrained minimization technique. The results are given in Section 7.4.

This type of model is primarily for use by groups concerned with future projects and advance planning of launch vehicles. It provides an initial attempt to obtain a valid and flexible tool that advance planners may use in the formative stages of design for the purpose of obtaining optimum design parameters. These design parameters are chosen to obtain the minimum over-all cost of the launch vehicle within the constraints of the performance required to carry out the mission. If the model were to be developed to a further level of sophistication, it would be necessary to make a specific determination of its use. Two widely varying examples of application would be use for obtaining detail design parameters and use solely as a guide for preliminary planning.

7.2 DEVELOPMENT OF COST FUNCTION

Definition of Variables

Total cost for developing, building, and launching a launch vehicle is to be expressed as a function of design variables. Only the costs for the launch vehicle will be considered, and the development of a total space vehicle cost function, including the spacecraft or other PL, is not attempted. In this model the launch vehicle is defined as the composite of the three booster stages and the instrument unit. The instrument unit is the guidance and control for all three booster stages. The PL or spacecraft is that package above the launch vehicle that the launch vehicle is to place in the desired orbit or space trajectory. The space vehicle is defined as the composite of the launch vehicle and the spacecraft or PL. The cost function expressing the cost of the total launch vehicle development and production program is the objective function of the optimization cost model.

In the construction of the cost model a typical launch vehicle is used as an example. A number of basic design assumptions must be made before the development of the model. These assumptions include the number of stages in the launch vehicle, the number of engines per stage, the type of propellant, and the total number of launch vehicles to be built. Additional assumptions

Development of Cost Function

are made when the constraint equations are considered, but the ones mentioned are of primary importance for the cost function.

For purposes of illustration the example is a three-stage launch vehicle. The first stage has five identical engines of the bell-shaped exhaust chamber type. The first-stage engines use liquid oxygen/rocket projectile (LOX/RP) propellant. The second stage also has five engines of the bell-shaped type. The second-stage engines also use LOX/RP propellant; however, they are assumed to be of a design and thrust different from those of the first-stage engines. The third stage has a single bell-shaped engine. The third-stage engine uses a liquid oxygen/liquid hydrogen (LOX/LH$_2$) propellant. An instrument unit stage is considered that contains the guidance and control systems of the individual stages. No PL is considered in the cost function, although it is in the constraint equations. Thus the model is strictly for the launch vehicle costs. The number of launch vehicles to be built and launched is designated as the constant K_1.

The primary costs of a launch vehicle program are incurred in three major areas: research and development (R&D), manufacture of hardware, and launch operations. The cost of facilities for manufacturing and launching, which could be a fourth major cost area, is not considered, and the use of existing facilities is assumed. The R&D and hardware costs are determined at the subsystem level for each stage and at the system level for the instrument unit. The launch operations cost is determined by an equation representing the entire launch vehicle.

The two major subsystems comprising each stage for both R&D and hardware costs are the airframe and propulsion subsystems. Other subsystems less significant in terms of over-all stage cost are not considered. The two subsystems considered are assumed to represent the entire stage. The equations representing the costs of the two subsystems are primarily in terms of weight and/or thrust variables. The CERs between the given variables and the cost are expressed by the equations for the different cost categories of airframe and propulsion.

In the development of the cost model, the variables are defined as follows:

X_{11} = Stage 1 airframe weight (thousands of pounds),
X_{12} = Stage 1 total inert weight (thousands of pounds),
X_{13} = Stage 1 mass fraction (dimensionless),
X_{14} = Stage 1 total thrust (thousands of pounds),
X_{15} = Stage 1 impulse propellant weight (thousands of pounds),
X_{16} = Stage 1 individual engine thrust (thousands of pounds),
X_{17} = Stage 1 length (feet),
X_{21} = Stage 2 airframe weight (thousands of pounds),
X_{22} = Stage 2 total inert weight (thousands of pounds),

X_{23} = Stage 2 mass fraction (dimensionless),
X_{24} = Stage 2 total thrust (thousands of pounds),
X_{25} = Stage 2 impulse propellant weight (thousands of pounds),
X_{26} = Stage 2 engine thrust (thousands of pounds),
X_{27} = Stage 2 length (feet),
X_{31} = Stage 3 airframe weight (thousands of pounds),
X_{32} = Stage 3 total inert weight (thousands of pounds),
X_{33} = Stage 3 mass fraction (dimensionless),
X_{34} = Stage 3 total thrust (thousands of pounds),
X_{35} = Stage 3 propellant weight (thousands of pounds),
X_{36} = Stage 3 impulse engine thrust (thousands of pounds),
X_{37} = Stage 3 length (feet),
X_{41} = Instrument unit weight (thousands of pounds),
t_1 = Stage 1 burn time (seconds),
t_2 = Stage 2 burn time (seconds),
t_3 = Stage 3 burn time (seconds).

Total Stage 1 R&D and Production Costs

The variables to be used in determining the Stage 1 cost function are defined above. This portion of the cost function, encompassing both R&D and production costs, is developed in terms of the airframe and propulsion subsystems. The Stage 1 design assumptions will be as indicated in the earlier description of the example launch vehicle.

Stage 1 Airframe R&D Cost

The first category of cost to be considered for Stage 1 is the airframe R&D cost. This cost category is a combination of airframe R&D design engineering, tooling, special test equipment, subsystems test hardware, and static test hardware, as well as subsystems and static test support and acceptance costs. The equation for the total airframe R&D cost as taken from the Lockheed study is

$$Y = 5272.77(X_{11})^{1.2781}(X_{12})^{-0.1959}(X_{13})^{2.4242}(X_{14})^{0.38745}(X_{15})^{-0.9904}.$$

The term Y equals the total Stage 1 airframe R&D cost in millions of dollars. The equation takes into consideration the three major design parameters of an airframe—weights (X_{11}, X_{12}, and X_{15}), thrust (X_{14}), and mass fraction (X_{13})—and is applicable to expendable chemical airframes. A detailed definition of the term "mass fraction" is given in Section 7.3. Also to be taken up in that section are the interrelationships of the variables. With regard to the variable propellant weight, X_{15}, it is noted that impulse propellant weight is equal to total propellant weight, and no residue subsequent to burning is assumed.

Stage 1 Engine R&D Cost

The next cost to be considered is that of the propulsion subsystem R&D cost. This cost category includes R&D engineering, test hardware, special test equipment, qualification testing, and propellant costs, through qualification. The equation for Stage 1 total R&D cost of the engine through qualification as taken from the Lockheed study is

$$Y = -247.963 + 160.909 \left(\frac{X_{16}}{10^3}\right)^{-0.146} + 282.874 \left(\frac{X_{16}}{10^3}\right)^{0.648}.$$

In this equation Y is equal to the total Stage 1 engine R&D cost through qualification in millions of dollars. The equation is for the specific example of Stage 1; that is, an engine burning LOX/RP with a bell-shaped nozzle. If the engine were using another type of propellant, the constants of the above equation would be changed, as is done in the case of the Stage 3 engine. When a different type of nozzle or a nuclear or air-breathing engine is used, a new equation must be determined.

The first stage is ignited on the earth's surface, and therefore engine sea-level thrust is used for the engine thrust variable X_{16}. As will be the case for Stages 2 and 3, engine-rated thrust is used for all upper stages.

Stage 1 Airframe Production Cost

The first subsystem to be considered in obtaining the Stage 1 production cost is the airframe. The basic equation for airframe production cost is given in terms of the cost of the first production unit and is taken from the Lockheed study. The airframe as defined here includes structure tanks, insulation, thrust structure, fairings, engine accessories, fuel systems, oxidizer system, stage controls, telemetry structure, pneumatic system, separation system, and the interstage (that portion attached to the stage at separation).

To obtain the total production cost of all airframe subsystems produced for Stage 1, the number produced is raised to a particular exponent, depending on the slope of the learning curve, and multiplied by the first production unit basic equation. This is an approximation commonly used for learning curves. All learning curve values used are taken from the Lockheed study, along with the applicable cost term. The number of airframes produced is assumed to be equal to the number of stages required and thus equal to the number of launch vehicles to be built and launched, K_1. The first-unit airframe production cost for Stage 1, in dollars, is

$$Y = 185{,}214(X_{11})^{0.3322}(X_{13})^{-1.5935}(X_{15})^{0.2362}(X_{17})^{0.1079}(K_2)^{0.1616},$$

where K_2 = number of engines for Stage 1 = 5.

The total airframe production cost for Stage 1, in millions of dollars, is

$$Y = 185{,}214(X_{11})^{0.3322}(X_{13})^{-1.5935}(X_{15})^{0.2362}(X_{17})^{0.1079}(5)^{0.1616}(K_1)^\gamma(10^{-6}),$$

where γ = learning curve slope = 0.90.

Stage 1 Engine Production Cost

The hardware cost of the propulsion subsystem is the next to be considered, and, as in the case of airframe production costs, the first production unit cost is determined. The first production unit cost equation, in millions of dollars, for a Stage 1 LOX/RP engine with a bell-shaped nozzle as taken from the Lockheed study, is

$$Y = 0.2085\left(\frac{X_{16}}{10^3}\right) + 2.509\left(\frac{X_{16}}{10^3}\right)^{0.736} + 0.9744\left(\frac{X_{16}}{10^3}\right)^{-0.229}.$$

The total engine production cost for Stage 1 engines, in millions of dollars, is

$$Y = \left[0.2085\left(\frac{X_{16}}{10^3}\right) + 2.509\left(\frac{X_{16}}{10^3}\right)^{0.736} + 0.9744\left(\frac{X_{16}}{10^3}\right)^{-0.229}\right](5K_1)^{0.93}.$$

The term $5K_1^\gamma$ is the total number of engines produced, raised to the exponent of the slope of the learning curve. For the given engine the slope of the learning curve equals 0.93. The equation is for the total cost of the production hardware, excluding spares.

Total Stages 2 and 3 R&D and Production Costs

The Stages 2 and 3 cost functions are developed in the same manner as those for Stage 1. The same subsystems are used, and the variables and design assumptions are as defined earlier in this section for the example vehicle. Since the Stage 2 design assumptions for propellant and number of engines are the same as those for Stage 1, the incorporation of Stage 2 variables is the only change to the cost function for Stage 2. Because of this it is not necessary to repeat the discussion for the development of the Stage 2 cost function.

The Stage 3 cost function, although developed in the same manner, has some different values, owing to the use of a LOX/LH$_2$ propellant and only one engine. This primarily affects the engine costs. The airframe costs differ only by the incorporation of Stage 3 variables and by a value of 1 for the number-of-engines term in the airframe production cost. Because of the difference in propellant, the Stage 3 engine R&D and production costs are discussed.

Development of Cost Function

Stage 3 Engine R&D Cost

The engine R&D cost for Stage 3 is determined from the same basic equation as that for Stages 1 and 2. The constants of the equation are changed because of the change in propellant from LOX/RP to LOX/LH$_2$ in Stage 3. The total R&D cost through qualification for the Stage 3 engine, in millions of dollars, is

$$Y = 32.591 + 181.806 \left(\frac{X_{36}}{10^3}\right)^{0.539} + 232.57 \left(\frac{X_{36}}{10^3}\right)^{0.772}.$$

This equation for a LOX/LH$_2$ engine is taken from the Lockheed study. The comments on the engine R&D cost for Stage 1 are applicable.

Stage 3 Engine Production Cost

The engine production cost for Stage 3 is determined from the same basic equation as for Stages 1 and 2. The constants of the equation are changed to account for the LOX/LH$_2$ propellant of Stage 3. The Stage 3 engine first production unit cost, in dollars, is

$$Y = \left[0.0705 \left(\frac{X_{36}}{10^2}\right) - 0.1807 \left(\frac{X_{36}}{10^2}\right)^{-1.33} + 166.87 \left(\frac{X_{36}}{10^2}\right)^{0.498}\right](10^5).$$

The total engine production cost for Stage 3 engines, in millions of dollars, is

$$Y = \left[0.0705 \left(\frac{X_{36}}{10^2}\right) - 0.1807 \left(\frac{X_{36}}{10^2}\right)^{-1.33} + 166.87 \left(\frac{X_{36}}{10^2}\right)^{0.498}\right](10^{-1})(K_1)^{0.93}.$$

The comments on Stage 1 engine production cost are applicable to Stage 3 engine production cost. The above equation for production unit cost was taken from the Lockheed study and is only for LOX/LH$_2$ engines.

This concludes the Stage 3 costs and the development by stage of R&D and hardware costs. The cost of the instrument unit, which is not considered as a stage, is determined in much the same manner in the following subsection.

Launch Vehicle Instrument Unit Cost

The guidance and control of the launch vehicle are provided to all three stages from a single instrument unit throughout the entire powered flight. The instrument unit is mounted on top of the third stage and beneath the PL. Thus 1 lb of instrument unit weight is equivalent to 1 lb of PL weight with respect to launch vehicle performance. The costs are determined by R&D and production equations, as was done for each stage.

Instrument Unit R&D Cost

The R&D cost of the instrument unit is given by an equation representing total R&D costs of the guidance and control system. This cost category includes R&D engineering, tooling, special test equipment, components, test hardware, and components integration. The equation for guidance and control R&D cost as taken from the Lockheed study is

$$Y = 10.35\{15,822[(X_{41})(10^3)]^{0.736}(10^{-6})\} - 35.5.$$

The variable X_{41} is the weight of the instrument unit in thousands of pounds. The equation does not include any flight-test hardware units. The cost of flight-test units could be determined using the single-unit cost, which is the bracketed portion of the R&D equation above. It is assumed for the purposes of the example that any flight testing of the instrument unit would involve the entire launch vehicle. Therefore flight-test hardware cost may be treated as a production hardware cost and no addition is required for the R&D cost equation. The above equation is the total instrument unit R&D cost in millions of dollars.

Instrument Unit Production Cost

The guidance and control system production cost is the last hardware cost to be considered. The equation for the instrument unit first-production unit cost in dollars as taken from the Lockheed study is

$$Y = 15,822[(X_{41})(10^3)]^{0.786}.$$

The total instrument unit production cost for all first stages, in millions of dollars, is

$$Y = \{15,822[(X_{41})(10^3)]^{0.786}\}(10^{-6})(K_1)^{\gamma}.$$

The learning curve slope γ equals 0.90 for the given equation and example. The total instrument unit hardware includes computer, adapter, platform, flight controls, and miscellaneous equipment.

This concludes the development of R&D and hardware costs for the example launch vehicle.

Launch Vehicle Operations Cost

The last major cost area to be considered is the launch operations cost. This cost is determined by an equation developed for the total launch vehicle, not by stages and subsystems. The operations cost is defined to include the costs of pad operations, propellant, transportation, and other operations support. The equation for this function as taken from the Lockheed study [1] is

$$Y = 8.5 \left[\frac{(X_{15} + X_{25} + X_{35})(C)^{0.460}}{1000} \right](K_1),$$

Development of Cost Function

where C = the number of stages = 3. This equation gives the total flight operations cost for all vehicles launched (K_1), in millions of dollars.

Total Cost Function

The total launch vehicle program cost is the sum of all the costs described in this section. The total cost, which is the objective function of the cost model, can be written as

Total cost = R&D cost + hardware cost + operating cost.

The first two elements of the total cost, R&D cost and hardware cost, were developed by individual stage and the instrument unit.

The first-stage costs are

Airframe R&D + LOX/RP propulsion R&D + airframe production unit (K_1) + LOX/RP engine production unit (5) (K_1).

The second-stage costs are

Airframe R&D + LOX/RP propulsion R&D + airframe production unit (K_1) + LOX/RP engine production unit (5) (K_1).

The third-stage costs are

Airframe R&D + LOX/LH$_2$ propulsion R&D + airframe production unit (K_1) + LOX/LH$_2$ engine production unit (1) (K_1).

The instrument unit costs are

Instrument unit R&D + instrument unit production unit (K_1).

The launch vehicle operating cost is

Launch operations (K_1).

Thus the complete objective function of the model to be minimized is as follows.

Stage 1

$$[5272.77(X_{11})^{1.2781}(X_{12})^{-0.1959}(X_{13})^{2.4242}(X_{14})^{0.38745}(X_{15})^{-0.9904}]$$

$$+ \left[-247.963 + 160.909\left(\frac{X_{16}}{10^3}\right)^{-0.146} + 282.874\left(\frac{X_{16}}{10^3}\right)^{0.648}\right]$$

$$+ [185,214(X_{11})^{0.3322}(X_{13})^{-1.5935}(X_{15})^{0.2362}(X_{17})^{0.1079}(5)^{0.1616}(K_1)^{0.90}(10)^{-6}]$$

$$+ \left[0.2085\left(\frac{X_{16}}{10^3}\right) + 2.509\left(\frac{X_{16}}{10^3}\right)^{0.736} + 0.9744\left(\frac{X_{16}}{10^3}\right)^{-0.229}\right](5K_1)^{0.93}$$

Stage 2

$$+ [5272.77(X_{21})^{1.2781}(X_{22})^{-0.1959}(X_{23})^{2.4242}(X_{24})^{0.38745}(X_{25})^{-0.9904}]$$

$$+ \left[-247.963 + 160.909 \left(\frac{X_{26}}{10^3}\right)^{-0.146} + 282.874 \left(\frac{X_{26}}{10^3}\right)^{0.648} \right]$$

$$+ [185,214(X_{21})^{0.3322}(X_{23})^{-1.5935}(X_{25})^{0.2362}(X_{27})^{0.1079}(5)^{0.1616}(K_1)^{0.90}(10)^{-6}]$$

$$+ \left[0.2085 \left(\frac{X_{26}}{10^3}\right) + 2.509 \left(\frac{X_{26}}{10^3}\right)^{0.736} + 0.9744 \left(\frac{X_{26}}{10^3}\right)^{-0.229} \right] (5K_1)^{0.93}$$

Stage 3

$$+ [5272.77(X_{31})^{1.2781}(X_{32})^{-0.1959}(X_{33})^{2.4242}(X_{34})^{0.38745}(X_{35})^{-0.9904}]$$

$$+ \left[32.591 + 181.806 \left(\frac{X_{36}}{10^3}\right)^{0.539} + 232.57 \left(\frac{X_{36}}{10^3}\right)^{0.772} \right]$$

$$+ [185,214(X_{31})^{0.3322}(X_{33})^{-1.5935}(X_{35})^{0.2362}(X_{37})^{0.1079}(1)^{0.1616}(K_1)^{0.90}(10)^{-6}]$$

$$+ \left[0.0705 \left(\frac{X_{36}}{10^2}\right) - 0.1807 \left(\frac{X_{36}}{10^2}\right)^{-1.33} + 166.87 \left(\frac{X_{36}}{10^2}\right)^{0.498} \right] (10)^{-1}(K_1)^{0.93}$$

Instrument Unit

$$+ \{10.35[15,822(X_{41}\cdot 10^3)^{0.786}(10)^{-6}] - 35.5\}$$
$$+ [15,822(X_{41}\cdot 10^3)^{0.786}(10)^{-6}(K_1)^{0.90}]$$

Launch Operations

$$+ 8.5 \left[\frac{(X_{15} + X_{25} + X_{35})(3)}{1000} \right]^{0.460} (K_1).$$

It should be noted that in the fourth term of the Stage 3 cost the coefficient 166.87 appears. In our numerical solutions discussed in Section 7.4 we mistakenly used the coefficient 16.687.

7.3 DEVELOPMENT OF CONSTRAINTS

Definition of Types of Constraints

It is the purpose of the constraints in the model to set forth the basic conditions of the problem and the limitations on each of the variables. The conditions and limitations on the given variables are determined by three sets of relationships. The definitional interrelationships of the seven variables of each stage are determined by the definition of the variables. The basic performance relationships are determined by the fundamental scientific laws and definitions of rocket propulsion. The third set of relationships is those

Development of Constraints

set up to express desired performance parameters. This section expresses as equalities or inequalities the required constraints from each of the three basic relationships for each of the three stages and for the instrument unit of the example model.

Total Stage 1 Constraints

The Stage 1 constraints are developed from the three basic types of relationships mentioned above. This is accomplished for Stage 1 in detail. The basic design assumptions of the example model, as used in the development of this objective function, are assumed to hold in this development of the constraint equations. Necessary additional assumptions are discussed as the constraints are developed.

Stage 1 Definitional Interrelationships

The definitional interrelationships result from definitions that allow the expression of certain variables in terms of other variables within the basic set of seven Stage 1 variables that appear in the objective function. Because of these definitional interrelationships it is possible to reduce the number of variables used in the statement of the problem. This could be accomplished by using the definitional interrelationships and eliminating those variables that are expressible in terms of other variables. It is not desirable to eliminate the redundant variables, however, because of the additional clarity in changing relationships from problem to problem, and because of the conciseness achieved by using an additional variable term.

The first definitional interrelationship of Stage 1 is the *determination of weights*. The equation is

$$(0.5)X_{12} = X_{11}.$$

This equation indicates that the airframe weight is equal to half the total stage inert weight for Stage 1. The other half of the total stage weight empty is in the form of the propulsion subsystem and equipment from other subsystems, such as instrumentation. The propulsion subsystem makes up a large percentage of the remaining half, with miscellaneous equipment being a much smaller part. The use of one-half as the airframe portion of the total stage weight empty is an arbitrary judgment based on values from similar stages. A curve to depict this fractional value could be generated based on the number of engines and the relative size of the airframe. However, this is not a useful effort for our purposes. Arbitrary judgment based on experience will again be used to determine values for Stages 2 and 3.

The second definitional interrelationship is that of the *stage thrust and the thrust of a single engine*. This equation for Stage 1 with its five engines shows that stage thrust is equal to the single-engine thrust multiplied by the number of engines. The equation is stated as

$$X_{14} = (5)X_{16}.$$

The number of engines, namely, five for Stage 1, is a basic design assumption given in Section 7.2.

The next definitional interrelationship is that of the dimensionless fractional value, *the stage mass fraction*. For Stage 1 this important parameter in airframe design is the ratio of the launch vehicle weight subsequent to Stage 1 burn and prior to Stage 1 separation to the initial launch vehicle weight.

This is expressed in equation form as

$$X_{13} = \frac{X_{12} + X_{22} + X_{25} + X_{32} + X_{35} + X_{41} + PL}{X_{12} + X_{15} + X_{22} + X_{25} + X_{32} + X_{35} + X_{41} + PL}.$$

More will be said of the stage mass fraction in the basic performance parameter section.

To maintain the *structural integrity* it is necessary to provide a relationship between the stage's inert weight and the propellant weight. This relationship is

$$12X_{12} \leq X_{15} \leq 16X_{12}.$$

These two inequalities state that the propellant weight for Stage 1 is between 12 and 16 times the Stage 1 inert weight. The range of 12 to 16 was arbitrarily selected on the basis of the design of similar stages. Thus the size of the stage and the amount of propellant it will hold are correlated.

The last variable to be considered for Stage 1 is the *stage length*, X_{17}. Some relationship between the length of the stage and the inert stage weight or propellant weight could probably be determined using volume and density constraints. It appears most appropriate, however, to let this variable fluctuate between arbitrary limits set because of the capability of handling or practicality of design. Thus

$$C_{17}^{l} \leq X_{17} \leq C_{17}^{\mu}.$$

The bounds C_{17}^{l} and C_{17}^{μ} are determined by selecting values from knowledge of the typical kind of stage required to meet the performance requirements. Thus the bounds are

$$125 \leq X_{17} \leq 150.$$

This completes the definitional interrelationships of Stage 1.

Stage 1 Basic Performance Relationships

The basic performance relationships are essentially the definitions of the basic relations of performance parameters used in rockets. These basic performance parameters are expressed in terms of the variables of the

Development of Constraints

example model. The parameters to be discussed are the more important performance parameters used in rockets.

One of the most important performance parameters is the *specific impulse*, I_s. This may be defined as the "impulse delivered per unit weight of propellant."

The equation for specific impulse in terms of the variables of the model is

$$I_s = \frac{F}{W} = \frac{(X_{14})(t_1)}{X_{15}}.$$

In dimensions, specific impulse is pounds of thrust divided by pounds per second of propellant flow. Thus the above equation gives the thrust divided by the propellant flow rate. For the LOX/RP propellant of Stage 1 the specific impulse can take on a range of values owing to chamber pressure, mixture ratio, exit pressure and velocity, and other factors. For the specific impulse of Stage 1 engines, bounds are given that are consistent with the type of propellant and desired performance. Thus the specific impulse equation is rewritten as a constraint as follows:

$$240 \leq \frac{(X_4)(t_1)}{X_{15}} \leq 290.$$

The next performance parameter to be discussed is the *stage mass fraction*, X_{13}. This design parameter, which is a variable in the model, is the ratio of the final vehicle weight after first-stage propellant burn to the initial vehicle weight, including propellant. This is expressed for Stage 1 by

$$\text{SMF}_1 = \frac{M_f}{M_0} = X_{13}.$$

In the development of the constraints of this model the difference between the initial and final weight is assumed to be the total propellant weight. This assumes that all propellant is burned and that there is no residual. This is done for simplification of the model and has little impact on the accuracy, as the residual is a very small percentage of the total propellant. If a greater degree of accuracy were necessary, residuals could be included in the model. The mass fraction is an important performance parameter since it has a major effect on stage velocity attainable. This can be seen in the later equations in the next subsection, where mass fraction is used. The variable X_{13} is bounded by typical values attainable through good stage design and necessary for the performance required in the model. Thus the Stage 1 mass fraction constraint is

$$0.25 \leq X_{13} \leq 0.30.$$

The *thrust-to-initial-weight* ratio for the stage is the next performance parameter to be considered. This ratio is self-explanatory and is stated as

$$\frac{F}{W_0} = \frac{X_{14}}{X_{12} + X_{15} + X_{22} + X_{25} + X_{32} + X_{35} + X_{41} + PL}.$$

The thrust is the total stage thrust X_{14}, and the weight is the initial launch vehicle weight plus propellant and PL. The bounds set on this parameter are typical of states required in the model performance range. Thus the thrust-to-weight ratio constraints are

$$1.2 \leq \frac{X_{14}}{X_{12} + X_{15} + X_{22} + X_{25} + X_{32} + X_{35} + X_{41} + PL} \leq 1.4.$$

This concludes the basic performance relationships subsection.

Stage 1 Desired Performance Parameter Relationships

The desired performance parameter relationships are determined from the basic equations of motion, and relate velocity and altitude to the given variables. Only velocity is discussed here, and the equation is for a simplified vertical trajectory with additional assumptions that simplify the model. The equation was chosen because of its simplicity and is intended to illustrate the incorporation of the variables into performance equations. More detailed equations used for detail design could also be adapted for use with the variables given in the objective function. The impact of the assumptions made is discussed in this subsection and in the subsection "Total Launch Vehicle Constraints."

The desired performance parameter for a simplified vertical trajectory is the velocity that can be achieved for a vertically ascending rocket assuming that (a) the earth is stationary, (b) the direction of thrust coincides with the flight path, and (c) no side forces are present. These assumptions basically mean that the effect of centrifugal and Coriolis forces and the effect of aerodynamic side loads are neglected. These effects are of relatively small magnitude. By definition of the trajectory, no maneuvering of the launch vehicle during the boost phase will be required.

Under the foregoing assumptions, the velocity of the total vehicle at the completion of the Stage 1 burn time, t_1, is given by

$$v_1 = \bar{c} \ln \left(\frac{M_0}{M_f}\right) - \bar{g} t_1 - \frac{(B_1 C_D A)}{W_0} + v_0.$$

Reviewing the terms of this equation, from last to first, the initial velocity v_0 at ignition of the first-stage engines is assumed to be zero, representing a

Development of Constraints

pad launch. The next term, $(B_1 C_D A)/W_0$, is the effect of the earth's atmosphere. This aerodynamic drag term is composed of the coefficient of drag, C_D, for the launch vehicle, the maximum cross-sectional area, A, of the launch vehicle, the initial weight, W_0, of the vehicle before ignition, and the term B_1. The term B_1 is represented by the following integral:

$$B_1 = g_0 \int_0^{t_1} \frac{\frac{1}{2}\rho v^2}{1 - \frac{\alpha t}{t_1}} \, dt.$$

In this model the aerodynamic drag will not be considered, and thus Stage 1 velocity will be greater than that which would actually be achieved. The penalty in burnout velocity due to atmospheric drag for a large, well-designed launch vehicle is less than 2 per cent. Thus the deletion is not important.

The next term, $\bar{g} t_1$, is the effect of gravity during the burn time, t_1. This term will not be considered in the velocity calculations, and thus the equation for velocity disregards all external forces and reduces to what can be called the "ideal velocity." To compensate for the deletion of the external forces the total velocity constraint will be in terms of an ideal velocity for the mission. This is discussed further in the last part of this section.

The remaining term of the velocity equation for Stage 1 represents the velocity imparted to the mass during the burn time t_1 under the idealized conditions reviewed above. Thus the velocity equation for Stage 1 is

$$v_1 = \bar{c} \ln \left(\frac{M_0}{M_f} \right),$$

where M_0 and M_f are respectively the initial mass of the total launch vehicle and the final mass subsequent to Stage 1 burn. The effective exhaust velocity, $\bar{c}$, is defined as the specific impulse, I_s, multiplied by the average gravity, $\bar{g}$.

The ratio M_0/M_f is the inverse of the stage mass fraction. Thus, when the expressions defined earlier for effective exhaust velocity and mass fraction are inserted, the velocity equation in terms of the variables of the model is

$$v_1 = \frac{X_{14} t_1 \bar{g}}{X_{15}} \ln \left(\frac{1}{X_{13}} \right).$$

Total Stages 2 and 3 Constraints

Stages 2 and 3 constraints are developed in the same manner as the Stage 1 constraints. The same three basic sets of relationships are developed as were used for Stage 1. The only changes to these constraints are the result of different design assumptions among the three stages and the incorporation of the applicable stage variables. Although the equality and inequality constraints for Stages 2 and 3 are not given until the end of this section, the following is a discussion of changes from Stage 1.

The determination of weight equations for Stages 2 and 3 differs by the fractional value of the airframe weight to the total stage inert weight. The arbitrary values for Stages 2 and 3 are 0.6 and 0.7, respectively.

The bounds on the inequality constraints for Stages 2 and 3 are changed from the Stage 1 bounds to be typical of good design values for middle and upper stages, respectively. The Stages 2 and 3 bounds are arbitrarily chosen in the same manner as those for Stage 1.

The variables in the terms for the stage mass fraction and the thrust-to-weight ratio constraints are in accordance with the sequence of stage separation and/or stage burning. Thus for Stages 2 and 3 the terms for these two constraints do not include the variables for the stage or stages already fired and separated. The variables in the terms for the other constraints contain only variables of the stage being described.

This concludes the discussion of Stages 2 and 3 constraints. The actual constraints used in the model are shown at the end of this section.

Instrument Unit Constraint

There is only one launch vehicle parameter of the instrument unit that appears in the objective function. This parameter, the instrument unit weight in thousands of pounds, must be constrained. Since no direct relation exists between instrument unit weight and the other variables and parameters that are considered in this model, the constraint is accomplished by an arbitrary bounding. Since the size and weight of the instrument unit depend on the sophistication required of the guidance and control system, the assumptions of the example model make this variable somewhat meaningless. Therefore the instrument unit constraint is not based on requirements of the example mission, but on the type of requirements that might be placed on a launch vehicle such as the example vehicle. Thus the variable is bounded by the following:

$$2.5 \leq X_{41} \leq 4.0.$$

Since the variable is in thousands of pounds, the bounds are also in thousands of pounds.

Total Launch Vehicle Constraints

The total launch vehicle constraints are the composite of the individual stage constraints. The three basic types of constraints ensure that each variable of the model is constrained and within the desired bounds. The definitional interrelationship constraints ensure that variables not used in the performance parameters or relations are constrained through their relationship to those variables that are used. The basic performance relationships ensure that the proper relationship is maintained between the design

Development of Constraints

of the individual stages. The desired performance parameters ensure that the total launch vehicle is capable of accomplishing the desired mission.

The total launch vehicle velocity is the summation of the incremental velocities of the individual stages. Thus the total launch vehicle velocity is given by

$$v_t = v_1 + v_2 + v_3.$$

This equation in terms of the variables of the model is given by

$$v_t = \left[\left(\frac{X_{14}t_1\bar{g}}{X_{15}}\right)\ln\left(\frac{1}{X_{13}}\right)\right] + \left[\left(\frac{X_{24}t_2\bar{g}}{X_{25}}\right)\ln\left(\frac{1}{X_{23}}\right)\right] + \left[\left(\frac{X_{34}t_3\bar{g}}{X_{35}}\right)\ln\left(\frac{1}{X_{33}}\right)\right].$$

This equation gives the terminal velocity that can be attained by the vehicle in a gravity-free vacuum if all the propulsive energy of all stages is applied in the same direction. If the mission velocity requirement takes into consideration ideal velocity calculations, an earth escape velocity constraint is given by the following inequalities:

$$35{,}000 \leq v_t \leq 50{,}000.$$

This completes the velocity constraint of the model.

Summary of Constraints

All the constraints have been discussed. The constraints required for the model are summarized as follows:

Stage 1

$$(0.5)X_{12} = X_{11},$$

$$X_{14} = (5)X_{16},$$

$$X_{13} = \left(\frac{X_{12} + X_{22} + X_{25} + X_{32} + X_{35} + X_{41} + PL}{X_{12} + X_{15} + X_{22} + X_{25} + X_{32} + X_{35} + X_{41} + PL}\right),$$

$$12X_{12} \leq X_{15} \leq 16X_{12},$$

$$125 \leq X_{17} \leq 150,$$

$$240 \leq \frac{X_{14}t_1}{X_{15}} \leq 290,$$

$$0.25 \leq X_{13} \leq 0.30,$$

$$1.2 \leq \frac{X_{14}}{X_{12} + X_{15} + X_{22} + X_{25} + X_{32} + X_{35} + X_{41} + PL} \leq 1.4,$$

Stage 2

$$(0.6)X_{22} = X_{21},$$

$$X_{24} = (5)X_{26},$$

$$X_{23} = \left(\frac{X_{22} + X_{32} + X_{35} + X_{41} + PL}{X_{22} + X_{25} + X_{32} + X_{35} + X_{41} + PL}\right),$$

$$10X_{22} \leq X_{25} \leq 12X_{22},$$

$$75 \leq X_{27} \leq 100,$$

$$240 \leq \frac{X_{24}t_2}{X_{25}} \leq 290,$$

$$0.24 \leq X_{23} \leq 0.29,$$

$$0.6 \leq \frac{X_{24}}{X_{22} + X_{25} + X_{32} + X_{35} + X_{41} + PL} \leq 0.75,$$

Stage 3

$$(0.7)X_{32} = X_{31},$$

$$X_{34} = X_{36},$$

$$X_{33} = \left(\frac{X_{32} + X_{41} + PL}{X_{32} + X_{35} + X_{41} + PL}\right),$$

$$7X_{32} \leq X_{35} \leq 9X_{32},$$

$$50 \leq X_{37} \leq 70,$$

$$340 \leq \frac{X_{34}t_3}{X_{35}} \leq 375,$$

$$0.16 \leq X_{33} \leq 0.21,$$

$$0.7 \leq \frac{X_{34}}{X_{32} + X_{35} + X_{41} + PL} \leq 0.9,$$

Instrument Unit

$$2.5 \leq X_{41} \leq 4.0,$$

Total Launch Vehicle

$$35{,}000 \leq \left[\left(\frac{X_{14}t_1\bar{g}}{X_{15}}\right)\ln\left(\frac{1}{X_{13}}\right)\right] + \left[\left(\frac{X_{24}t_2\bar{g}}{X_{25}}\right)\ln\left(\frac{1}{X_{23}}\right)\right]$$

$$+ \left[\left(\frac{X_{34}t_3\bar{g}}{X_{35}}\right)\ln\left(\frac{1}{X_{33}}\right)\right] \leq 50{,}000.$$

7.4 APPLICATION OF MODEL

Method of Solution and Results

The nonlinear programming problem has been solved using the sequential unconstrained minimization technique.

The values shown in Table 7.1 for Model I are with the objective function and constraints as developed in Sections 7.2 and 7.3. The objective function parameter K_1 was given a value of 10, and the PL was assumed to be 20,000 lb. Thus the program consisted of the development, building, and launching of 10 launch vehicles. The values of the objective function computed for starting point sets a and b were \$3.36 billion and \$3.16 billion, respectively. The value of the objective function computed at the optimal solution to Model I was \$2.53 billion. The same answer (optimal solution) was obtained for both starting point sets.

Table 7.1 Starting and Solution Values of the Variables for Models I and II

Variable	Starting Point Set a	Starting Point Set b	Solution for Model I	Solution for Model II
X_{11}	150	100	68	54
X_{12}	300	200	136	109
X_{13}	0.28	0.29	0.3	0.3
X_{14}	7000	5500	3733	3353
X_{15}	4000	3000	2177	1956
X_{16}	1400	1100	746	671
X_{17}	135	150	125	125
X_{21}	54	39	28	24
X_{22}	90	65	47	40
X_{23}	0.27	0.286	0.29	0.29
X_{24}	800	700	478	438
X_{25}	900	750	566	518
X_{26}	160	140	96	88
X_{27}	85	100	75	75
X_{31}	17.5	16	11	9.5
X_{32}	25	23	16	13.6
X_{33}	0.19	0.20	0.21	0.21
X_{34}	200	180	129	120
X_{35}	200	190	145	136
X_{36}	200	180	129	120
X_{37}	60	70	50	50
X_{41}	3	4	2.5	2.5
t_1	150	145	155	155
t_2	300	275	314	314
t_3	350	370	403	403
Total cost (billions)	\$3.36	\$3.16	\$2.53	\$2.33

A variation of Model I (called Model II) was solved using different bounds for the structural integrity constraint. The bounds of the propellant-to-stage-weight multiple were changed for Stage 1 from 12 and 16 to 10 and 18. For Stage 2 the bounds were changed from 8 and 12 to 7 and 13, and for Stage 3 from 7 and 9 to 6 and 10. The same starting point sets were used as in Model I, and the solution values are also given in Table 7.1. The value of the objective function computed at the solution of Model II was $2.33 billion. The same optimal solution was obtained for both starting point sets.

Analysis of Results

One of the most difficult problems in nonlinear programming is dealing with local and global minima. No algorithm has been developed that assures convergence to the global minimum for a complicated, nonlinear, nonconvex programming problem such as the model developed in this chapter. However, by using various starting values, convergence to the same optimal solution indicates that a global minimum may have been achieved. Examination of the results of the computer runs for Models I and II has shown that the solution values for the variables are nearly identical for both starting point sets. Obtaining the same results for two sets of starting points over a number of runs indicates that the values of the objective function obtained for the two models may be the global minima.

The solution value of the stage mass fraction is interesting to note. The solution values for all three stages in both variations of the model attain their upper bounds. This indicates that, although a lower mass fraction is desirable from a good engineering design standpoint, the cost of achieving it is greater than the performance benefit gained. This is at least true within the constraints of the model and can intuitively be said to be true in areas of missile design. This can be shown, practically speaking, in that it is clear that beyond a certain point refinements in design are no longer practical, and simply a larger over-all vehicle is required.

When the bounds of the structural integrity constraint for the second variation of the model were changed, the variables again attained their upper bound for all three stages. When the structural integrity constraint is deleted, the stage structural weight goes toward zero, and the propellant weight goes as high as the remaining constraints will allow. Thus the propellant-to-stage-weight ratio will go as high as the structural integrity constraint will allow. This indicates that a structural complexity factor to include the greater design sophistication of a stage with a high propellant-to-stage-weight ratio could profitably be included in the objective function.

The thrust-to-weight ratio attained its lower bound for all three stages and both variations of the model. Analysis of this result is the same as that given for the stage mass fraction. The greater the thrust-to-weight ratio, the

Application of Model

greater the cost, and within the limits of the model the increase in performance capability was not required.

The specific impulse constraint for Stages 1 and 2 remained the same between the two variations of the model and between the two stages. It is expected that this constraint would respond similarly for the two stages, as they utilize the same fuel; however, it is not expected that they would remain the same. It is possible that the model is overconstrained in this area or that this may be a reflection of an optimum point in the CERs, which are the same for the two stages. Which of these possibilities is in fact the case was not determined.

The velocity constraint was not pushed to its lower bound in either variation of the model. In both models the velocity constraint attained the value of 38,632 ft/sec. This indicates that a relaxing of the constraints is required to obtain the optimum-sized vehicle to accomplish the minimum velocity objectives. This could be achieved by deletion of the thrust-to-weight ratio constraints, which, along with the burn times, determine the acceleration. In relaxing some of the constraints to attain a minimum velocity objective, consideration must be given to retaining a sound launch vehicle design and not allowing basic design parameters to achieve infeasible values.

Since the stage length variables are unrelated to the performance constraints as utilized in the model, these variables may be adjusted after the computer run. Thus the stage length may be put in the proper relationship to the other stage design variables. The total cost function will be affected, and incorporation of the new length will result in a more accurate total cost value.

Solution of the nonlinear programming problem has demonstrated that this type of complex nonlinear model is computationally feasible. The analysis of the results has given some indication of the usefulness of the model, and further discussion in this area is given in the next subsection. The fact that the model developed is a prototype and that numerous improvements can be made should be emphasized. A discussion of some of these proposed improvements is given in a later subsection.

Model Sensitivity to Changes in Constraints

Associated with the solution of any nonlinear programming problem is a set of values called "shadow prices" or generalized Lagrange multipliers, one for each inequality constraint that represents the sensitivity of the value of the objective function to a tightening or relaxation of the constraints. For those constraints that are not binding at the solution, the corresponding shadow prices are equal to zero. The interesting shadow prices are those associated with the binding constraints.

The theory behind the use of shadow prices is that, if the original problem were changed and the ith constraint relaxed by an amount δ [that is,

Table 7.2 Sensitivity to Changes in Limits of Binding Constraints in Model I

Constraint	Unit Change	Savings in Total Cost (millions of dollars)
Stage mass fraction, upper bound		
Stage 1	+0.01	41.3
Stage 2	+0.01	50.0
Stage 3	+0.01	121.5
Thrust-to-weight ratio, lower bound		
Stage 1	−0.1	24.7
Stage 2	−0.1	13.1
Stage 3	−0.1	15.8

$g_i(x) + \delta \geq 0$], then the change in the optimum value of the objective function would be approximately equal to $-\delta \bar{\mu}_i$, where $\bar{\mu}_i$ is the shadow price associated with the ith constraint at the solution of the original problem.

The use of these values is particularly important in those design problems where one is interested in the sensitivity of the model to changes in the design requirements. This is illustrated by using the shadow prices produced by the sequential unconstrained minimization technique.

The two constraints to be considered are the stage mass fraction and thrust-to-weight ratio inequalities. The stage mass fraction constraint sought its upper bound for all three stages in both models. The thrust-to-weight constraint sought its lower bound for all three stages in both models. Table 7.2 shows the savings in millions of dollars that could be realized by allowing the stage mass fraction to go an additional hundredth higher than the upper bounds in the original model, and the savings for allowing the thrust-to-weight constraint to go an additional tenth lower than the lower bounds in the original model. Thus Table 7.2 shows the sensitivity of the objective function to unit changes in the limits of the binding constraints.

Possible Extensions of the Model

Two types of changes can be usefully accomplished to improve the results from the model. One type of change is a variation within the framework of the present model. The second type of change is the building of a new model utilizing the basic ideas of the prototype model.

One type of variation that has been accomplished and discussed is the changing of the values of the constraint bounds. Perturbations of the model such as this are useful in analyzing the results and in resolving some changes in design. Another variation that can be easily accomplished is changing the basic design assumptions, such as the number of stages or number of engines per stage. This could be useful in determining optimum staging. Numerous other changes within the framework of the prototype model are possible.

The second type of change would essentially require the building of a new model. This is desirable from several standpoints. New models could be developed with consideration of specific uses dictating their design. Also new models could incorporate the latest CERs and utilize more complex performance constraints as required. With the advent of more accurate CERs, more subsystems could be considered, and possibly to a lower level of detail than the subsystem. As more programming models are developed, undoubtedly many other areas of improvement will become evident.

Conceptual and Preliminary Design and Costing

The model presented in this paper is applicable to the conceptual and preliminary design phases in the development of a launch vehicle. Its usefulness is limited to those phases performed before detail program definition. In the first two phases, however, the model can serve as a flexible, quick-response planning tool.

Descriptive cost models are being used at present for future projects planning by NASA and in industry. These models are used for cost prediction, funding, and scheduling. Other programming models are used to a limited extent in engineering design during the conceptual and preliminary design phases. However, a model that enables consideration of both engineering design and costing through optimization is not in use. The subject model is a prototype development to fill this void.

The difficulties in attaining an effective cost model are similar whether it is of the descriptive or optimization type. Both are restricted by the degree of accuracy of CERs and seldom are able to go below the subsystem level. The optimization model has the additional difficulty of being very hard to solve through mathematical programming methods. The use of SUMT and future refinements to this or other algorithms promises to open the door for complicated total system programming models.

7.5 SOURCE OF PROBLEM AND REFERENCES

Source

The nonlinear programming model was developed by Benjamin C. Rush as a student of one of the authors at The George Washington University. His thesis [2] documents the research. We collaborated with him in programming and solving the model. The model is presented in an RAC technical paper subsequently published in *Operations Research* [3].

References

[1] Lockheed Missiles & Space Company, "Launch Vehicle Component Costs Study," Vol. II, Technical Report prepared for the National Aeronautics and Space Administration under Contract NAS 8-11368, Huntsville, Ala., 1965.

[2] B. C. Rush, "A Nonlinear Programming Model for Optimization of Launch Vehicle Design Parameters To Obtain a Minimum Cost," M.B.A. thesis, George Washington University, Washington, D.C., June 1966.
[3] B. C. Rush, J. Bracken, and G. P. McCormick, "A Nonlinear Programming Model for Launch Design and Costing," Technical Paper TP-231, Research Analysis Corporation, McLean, Va., September 1966; also published in *Operations Res.*, **15**, 185–210 (1967).
[4] H. S. Seifert (Ed.), *Space Technology*, Wiley, New York, 1959.
[5] G. P. Sutton, *Rocket Propulsion Elements: An Introduction to the Engineering of Rockets*, 3rd Ed., Wiley, New York, 1963.

8
PARAMETER ESTIMATION IN CURVE FITTING

In this chapter we discuss the application of nonlinear programming to the fitting of linear and nonlinear regression models. It is possible to expand remarkably the criteria employed in fitting linear regression models by using nonlinear programming. The nonlinear programming problems are in all cases convex. Nonlinear regression models are much more difficult, yielding nonconvex nonlinear programming problems, but the general approach is promising and is quite natural when there are constraints on the parameters.

8.1 LINEAR REGRESSION WITH VARIOUS CRITERIA

We consider the problem of choosing the coefficients of linear regression models by minimizing functions of vertical deviations. We define the vertical deviations (α_i's) by

$$y_i - (x_{i1}b_1 + \cdots + x_{in}b_n) = \alpha_i, \quad i = 1, \ldots, m, \tag{8.1}$$

where m is the number of observations, y_i is the ith observation on the dependent variable, $x_{i1}, \ldots, x_{in}$ are the ith observations on the n independent variables, and $b_1, \ldots, b_n$ are the n regression coefficients to be determined. We define weights on the observations by $w_1, \ldots, w_m$. The criteria for choosing the coefficients are as follows:

(a) minimize $\sum_{i=1}^{m} w_i |\alpha_i|^p$ for $p \geq 1$.

(b) minimize $\{\text{maximum } w_i |\alpha_i|, \quad i = 1, \ldots, m\}$.

Consideration of these criteria requires some discussion. Frequently in practice the assumptions of classical least squares are not met. For instance, a weighted least-squares procedure would be desirable for dealing with heteroscedasticity. More serious cases result in such problems as errors of measurement, dependency in the error terms, or mis-specified relations. Under these practical conditions one may search for a "robust" estimator by Monte Carlo methods. In these cases estimators defined as arbitrary integer powers of weighted residuals cover a wide spectrum of possibly useful estimators. Ashar and Wallace [1] provide an example with a sampling study of minimum absolute deviation estimators.

The mathematical programming approach allows considerable power and flexibility in constraining regression coefficients or deviations. A study by Meyer and Glauber [10] used linear programming with constraints on the coefficients and a criterion of minimizing the sum of the absolute deviations. Constrained regression problems are mentioned in econometrics (see Goldberger [6]), and ability to specify constraints would be useful in incorporating prior knowledge about parameters (Theil [11]).

Application of linear programming to minimizing the sum of absolute deviations and to minimizing the maximum absolute deviation has been discussed by Charnes, Cooper, and Ferguson [2], Kelley [9], Fisher [5], and Wagner [12]. Wagner [13] shows that minimizing the sum of the squared deviations can be written as a problem with linear constraints and a quadratic criterion function.

Minimizing Functions of Absolute Vertical Deviations

The ith vertical deviation is defined by (8.1), and we consider criterion (a). Denoting the ith absolute vertical deviation by β_i, we may write limitations on the deviations from below:

$$-y_i + (x_{i1}b_1 + \cdots + x_{in}b_n) + \beta_i \geq 0, \qquad i = 1, \ldots, m, \quad (8.2a)$$

and limitations on the deviations from above may be written

$$y_i - (x_{i1}b_1 + \cdots + x_{in}b_n) + \beta_i \geq 0, \qquad i = 1, \ldots, m. \quad (8.2b)$$

The final constraint is that the deviations be non-negative:

$$\beta_i \geq 0, \qquad i = 1, \ldots, m. \quad (8.2c)$$

The coefficients b_j ($j = 1, \ldots, n$) may assume any value unless otherwise restricted. Additional restrictions on the b_j's and β_i's may be included if desired. The mathematical programming problem is as follows. Choose $b_1, \ldots, b_n$ and $\beta_1, \ldots, \beta_m$ to

$$\text{minimize} \sum_{i=1}^{m} w_i \beta_i^p \quad (8.3)$$

subject to (8.2a), (8.2b), and (8.2c).

When $p = 1$ and $w_i = 1$ ($i = 1, \ldots, m$) the mathematical programming problem is the linear programming problem of minimizing the sum of the absolute deviations. When $p > 1$ it is a convex nonlinear programming problem.

The ith vertical deviation is defined by (8.1), and we consider criterion (b). Denoting the maximum absolute weighted deviation by ϵ, we may write limitations on this deviation from below and above:

$$w_i[y_i - (x_{i1}b_1 + \cdots + x_{in}b_n)] \leq \epsilon, \quad i = 1, \ldots, m,$$
$$w_i[(x_{i1}b_1 + \cdots + x_{in}b_n) - y_i] \leq \epsilon, \quad i = 1, \ldots, m. \quad (8.4)$$

Adding the constraint that ϵ must be non-negative, we obtain

$$-w_i y_i + w_i(x_{i1}b_1 + \cdots + x_{in}b_n) + \epsilon \geq 0, \quad i = 1, \ldots, m,$$
$$w_i y_i - w_i(x_{i1}b_1 + \cdots + x_{in}b_n) + \epsilon \geq 0, \quad i = 1, \ldots, m, \quad (8.5)$$

$$\epsilon \geq 0.$$

The coefficients $b_1, \ldots, b_n$ may assume any value unless otherwise restricted. The mathematical programming problem is as follows. Choose $b_1, \ldots, b_n$ and ϵ to

$$\text{minimize } \epsilon \quad (8.6)$$

subject to (8.5).

When the w_i's are all equal to 1, this is the Chebyshev criterion. In the linear regression model formulated here the constraints and objective function are both linear, and so this is a linear programming problem.

Minimizing Functions of Vertical Deviations

We give here two nonlinear optimization formulations of the linear regression problem for criterion (a) that are valid only for even positive p. The two formulations are equivalent, and the problems are convex.

The first is a nonlinear programming problem with m equality constraints and $n + m$ variables. Choose $b_1, \ldots, b_n$ and $\alpha_1, \ldots, \alpha_m$ to

$$\text{minimize } \sum_{i=1}^{m} w_i \alpha_i^p \quad (8.7)$$

subject to

$$-y_i + (x_{i1}b_1 + \cdots + x_{in}b_n) + \alpha_i = 0, \quad i = 1, \ldots, m. \quad (8.8)$$

The second nonlinear optimization problem is a direct minimization of the m term, n variable functions:

$$\sum_{i=1}^{m} w_i[y_i - (x_{i1}b_1 + \cdots + x_{in}b_n)]^p. \quad (8.9)$$

If constraints are imposed on the b_i's and/or deviations this is a nonlinear programming problem. Otherwise it is an unconstrained minimization problem.

When $p = 2$ this criterion is the least-squares criterion and the first formulation is a quadratic programming problem. As p increases the criterion approaches the Chebyshev criterion. The problem has been investigated for increasing p by Goldstein, Levine, and Hereshoff [7].

8.2 EXAMPLE OF A LINEAR REGRESSION PROBLEM WITH VARIOUS CRITERIA

In this section we present results of nonlinear programming solution of a linear regression problem involving one dependent variable and three independent variables, with the coefficient b_1 to be determined as an autonomous term. Data in Table 8.1 are 20 observations on cotton yarn characteristics, taken from Duncan [4]. A column of 1's for x_{i1} ($i = 1, \ldots, 20$) ensures the autonomous term.

We use the formulation for minimizing the sum of the weighted absolute vertical deviations to integer powers $p = 1, 2, 3, 4$. The weights are taken

Table 8.1 Data from Duncan [4], p. 517

i	y_i	x_{i1}	x_{i2}	x_{i3}	x_{i4}
1	99	1	85	76	44
2	93	1	82	78	42
3	99	1	75	73	42
4	97	1	74	72	44
5	90	1	76	73	43
6	96	1	74	69	46
7	93	1	73	69	46
8	130	1	96	80	36
9	118	1	93	78	36
10	88	1	70	73	37
11	89	1	82	71	46
12	93	1	80	72	45
13	94	1	77	76	42
14	75	1	67	76	50
15	84	1	82	70	48
16	91	1	76	76	41
17	100	1	74	78	31
18	98	1	71	80	29
19	101	1	70	83	39
20	80	1	64	79	38

Example of a Linear Regression Problem

Table 8.2 Computed Regression Coefficients

Objective Function	b_1	b_2	b_3	b_4	Value of Objective Function
$\sum_{i=1}^{m} \lvert \alpha_i \rvert$	7.563	1.187	.375	$-.843$	8.900
$\sum_{i=1}^{m} \alpha_i^2$	39.328	1.069	.164	$-.936$	69.399
$\sum_{i=1}^{m} \lvert \alpha_i \rvert^3$	42.013	1.033	.204	$-.988$	5,227.6
$\sum_{i=1}^{m} \alpha_i^4$	46.827	1.019	.188	-1.045	40,265.6
$\sum_{i=1}^{m} \left\lvert \frac{\alpha_i}{y_i} \right\rvert$	7.700	1.187	.373	$-.812$	.92410
$\sum_{i=1}^{m} \left(\frac{\alpha_i}{y_i} \right)^2$	53.756	1.006	.048	$-.975$	.07494
$\sum_{i=1}^{m} \left\lvert \frac{\alpha_i}{y_i} \right\rvert^3$	55.268	.957	.105	-1.008	.00608
$\sum_{i=1}^{m} \left(\frac{\alpha_i}{y_i} \right)^4$	60.308	.949	.088	-1.080	.00052

to be the reciprocals of the observed values y_i ($i = 1, \ldots, 20$). Table 8.2 presents the results of the computations.

The nonlinear programming model for the third case of minimizing $\sum_{i=1}^{m} \left\lvert \frac{\alpha_i}{y_i} \right\rvert^3$, for example, is as follows. Choose $b_1, \ldots, b_4$ and $\beta_1, \ldots, \beta_{20}$ to

$$\text{minimize} \quad \left(\frac{1}{99}\right)^3 \beta_1^3 + \cdots + \left(\frac{1}{80}\right)^3 \beta_{20}^3$$

subject to

$$-99 + (1 \cdot b_1 + 85 b_2 + 76 b_3 + 44 b_4) + \beta_1 \geq 0$$
$$\vdots$$
$$-80 + (1 \cdot b_1 + 64 b_2 + 79 b_3 + 84 b_4) + \beta_{20} \geq 0$$
$$99 - (1 \cdot b_1 + 85 b_2 + 76 b_3 + 44 b_4) + \beta_1 \geq 0$$
$$\vdots$$
$$80 - (1 \cdot b_1 + 64 b_2 + 79 b_3 + 38 b_4) + \beta_{20} \geq 0$$
$$\beta_1 \geq 0, \ldots, \beta_{20} \geq 0.$$

8.3 NONLINEAR REGRESSION WITH VARIOUS CRITERIA

In this section we consider the problem of choosing coefficients of nonlinear regression models by minimizing functions of vertical deviations. We define the vertical deviations (α_i's) by

$$y_i - f(x_{i1}, \ldots, x_{in}; b_1, \ldots, b_r) = \alpha_i, \qquad i = 1, \ldots, m, \qquad (8.10)$$

where m is the number of observations, y_i is the ith observation on the dependent variable, and the function $f(\cdot\,;\cdot)$ is a nonlinear function with r parameters $b_1, \ldots, b_r$. The weights are again denoted by $w_1, \ldots, w_m$.

Minimizing Functions of Absolute Vertical Deviations

Denoting the ith absolute deviation by β_i, we may write the following mathematical programming model for minimizing the sum of the weighted absolute deviations to a power equal to or greater than 1. Choose $b_1, \ldots, b_r$ and $\beta_1, \ldots, \beta_m$ to

$$\text{minimize} \sum_{i=1}^{m} w_i \beta_i^p \qquad (8.11)$$

subject to

$$-y_i + f(x_{i1}, \ldots, x_{in}; b_1, \ldots, b_r) + \beta_i \geq 0, \qquad i = 1, \ldots, m,$$
$$y_i - f(x_{i1}, \ldots, x_{in}; b_1, \ldots, b_r) + \beta_i \geq 0, \qquad i = 1, \ldots, m,$$
$$\beta_i \geq 0, \qquad i = 1, \ldots, m. \qquad (8.12)$$

Since $f(\cdot\,;\cdot)$ is a nonlinear function the constraints are always nonlinear. For $p = 1$ the objective function is linear, but for all $p > 1$ it is nonlinear. The objective function is always convex. However, the first and second sets of m constraints, taken together, cannot be concave, and so the nonlinear programming problem is not convex.

Some nonlinear regression problems have been solved using this formulation, one of which will be given as an example.

For minimizing the maximum weighted absolute deviation, denoted by ϵ, a nonlinear programming problem may be written as follows. Choose $b_1, \ldots, b_r$ and ϵ to

$$\text{minimize } \epsilon \qquad (8.13)$$

subject to

$$-w_i y_i + f(x_{i1}, \ldots, x_{in}; b_1, \ldots, b_r) + \epsilon \geq 0, \qquad i = 1, \ldots, m,$$
$$w_i y_i - f(x_{i1}, \ldots, x_{in}; b_1, \ldots, b_r) + \epsilon \geq 0, \qquad i = 1, \ldots, m,$$
$$\epsilon \geq 0. \qquad (8.14)$$

This is a nonconvex programming problem since the constraints are not concave.

Minimizing Functions of Vertical Deviations

As in the corresponding observation in linear regression with various criteria, two equivalent nonlinear optimization formulations valid only for even positive p may be given. But in the case of nonlinear regression the problems are not convex.

The first is a nonconvex nonlinear programming problem with m equality constraints and $r + m$ variables.

Choose $b_1, \ldots, b_r$ and $\alpha_1, \ldots, \alpha_m$ to

$$\text{minimize} \sum_{i=1}^{m} w_i \alpha_i^p \qquad (8.15)$$

subject to

$$-y_i + f(x_{i1}, \ldots, x_{in}; b_1, \ldots, b_r) + \alpha_i = 0, \qquad i = 1, \ldots, m. \qquad (8.16)$$

The second is a direct minimization of the m-term, r-variable nonconvex function,

$$\sum_{i=1}^{m} w_i [y_i - f(x_{i1}, \ldots, x_{in}; b_1, \ldots, b_r)]^p. \qquad (8.17)$$

If additional constraints are imposed, this becomes a nonconvex nonlinear programming problem.

8.4 EXAMPLE OF A NONLINEAR REGRESSION PROBLEM

We consider a nonlinear regression problem solved by Hartley [8] in an article on a procedure for fitting nonlinear regression functions by least squares using a modification of the Gauss-Newton method of iterative solution.

The nonlinear regression model is Mitcherlisch's law of diminishing returns,

$$y = b_1 + b_2 e^{b_3 x},$$

where y is the dependent variable, x the independent variable, and b_1, b_2, and b_3 the parameters to be determined. Hartley's data are as follows, with six observations on the variables.

Dependent Variable Observation y	Independent Variable Observation x
127	−5
151	−3
379	−1
421	5
460	3
426	1

We use the first nonlinear programming formulation for minimizing functions of absolute vertical deviations. The nonlinear programming problem is as follows.

Choose $b_1, \ldots, b_3, \beta_1, \ldots, \beta_6$ to minimize the nonlinear (quadratic) objective function

$$\beta_1^2 + \cdots + \beta_6^2$$

subject to the nonlinear constraints

$$-127 + b_1 + b_2 e^{b_3(-5)} + \beta_1 \geq 0,$$
$$-151 + b_1 + b_2 e^{b_3(-3)} + \beta_2 \geq 0,$$
$$-379 + b_1 + b_2 e^{b_3(-1)} + \beta_3 \geq 0,$$
$$-421 + b_1 + b_2 e^{b_3(1)} + \beta_4 \geq 0,$$
$$-460 + b_1 + b_2 e^{b_3(3)} + \beta_5 \geq 0,$$
$$-426 + b_1 + b_2 e^{b_3(5)} + \beta_6 \geq 0,$$
$$127 - [b_1 + b_2 e^{b_3(-5)}] + \beta_1 \geq 0,$$
$$151 - [b_1 + b_2 e^{b_3(-3)}] + \beta_2 \geq 0,$$
$$379 - [b_1 + b_2 e^{b_3(-1)}] + \beta_3 \geq 0,$$
$$421 - [b_1 + b_2 e^{b_3(1)}] + \beta_4 \geq 0,$$
$$460 - [b_1 + b_2 e^{b_3(3)}] + \beta_5 \geq 0,$$
$$426 - [b_1 + b_2 e^{b_3(5)}] + \beta_6 \geq 0,$$
$$\beta_1 \geq 0, \ldots, \beta_6 \geq 0.$$

Results obtained from solving the nonlinear programming problem agree with Hartley's and are as follows:

$$b_1 = 523.3, \qquad b_2 = -156.9, \qquad b_3 = -.1997.$$

8.5 MAXIMUM LIKELIHOOD ESTIMATION

Another common method of finding estimates of parameter values that explain observed data is the method of maximum likelihood estimation. Let $b = (b_1, \ldots, b_r)$ be r unknown parameters of a frequency function $g_0(y, b)$ of the random variable y. Let $y_1, \ldots, y_m$ be m observations of y. The *likelihood function* associated with these observations and frequency function is

$$L(y_1, \ldots, y_m, b_1, \ldots, b_r) = g_0(y_1, b) g_0(y_2, b) \cdots g_0(y_m, b). \quad (8.18)$$

Maximum Likelihood Estimation

An important method of estimating the values of $b_1, \ldots, b_r$ based on the observed $y_1, \ldots, y_m$ is to maximize the likelihood function (8.18).

The $\bar{b}_1, \ldots, \bar{b}_r$ that maximize (8.18) are called maximum likelihood estimators and have many desirable statistical properties. The reader is referred to Cramér [3] for a fuller discussion.

The problem of maximizing $L(y_1, \ldots, y_m, b)$ is an unconstrained mathematical programming problem. Sometimes it is also desirable to constrain the b_i's so that certain physical requirements are not violated. This is the case for the maximum likelihood problem discussed next. In such a case a constrained mathematical programming problem results.

Since the logarithm of the likelihood function achieves its maximum at the same $\bar{b}$ as the likelihood function itself, the general problem of likelihood estimation is stated as follows.

Find values $(\bar{b}_1, \ldots, \bar{b}_r)$ that

$$\text{maximize} \sum_{i=1} \ln g_0(y, b) \tag{8.19}$$

subject to

$$g_i(b) \geq 0, \quad i = 1, \ldots, l. \tag{8.20}$$

Our example comes from the biomedical area. It is hypothesized that the population of systolic blood pressures can be separated into three separate groups. The distribution of blood pressures within each of these groups can be represented by a normal frequency function. Let p_1, p_2, and p_3 represent the proportions of the population in each of the three groups. Let (μ_1, σ_1), (μ_2, σ_2), and (μ_3, σ_3) be the means and standard deviations of the normal frequency functions corresponding to each group. These nine values correspond to the unknown $\{b_j\}$ parameters.

Then under these assumptions the frequency function for the random variable y, which denotes systolic blood pressure, is obtained by summing the frequency functions of the individual groups times their probability of occurrence to yield

$$\frac{1}{\sqrt{2\pi}} \sum_{k=1}^{3} \frac{p_k}{\sigma_k} \exp\left[-\frac{(y - \mu_k)^2}{2\sigma_k^2}\right], \tag{8.21}$$

where

$$p_1 + p_2 + p_3 = 1.$$

There are eight parameters in this frequency function since one proportion, or probability, can be eliminated. Let $p_3 = 1 - p_1 - p_2$. Using the general problem of maximum likelihood estimation given by (8.19) and (8.20), the mathematical programming problem is as follows.

Find values of $(\bar{p}_1, \bar{p}_2, \bar{\mu}_1, \bar{\mu}_2, \bar{\mu}_3, \bar{\sigma}_1, \bar{\sigma}_2, \bar{\sigma}_3)$ that

$$\text{maximize} \sum_{i=1}^{m} \ln\left(\frac{1}{\sqrt{2\pi}}\left\{\frac{p_1}{\sigma_1}\exp\left[-\frac{(y_i - \mu_1)^2}{2\sigma_1^2}\right] + \frac{p_2}{\sigma_2}\exp\left[-\frac{(y_i - \mu_2)^2}{2\sigma_2^2}\right] \right.\right.$$
$$\left.\left. + \frac{1 - p_1 - p_2}{\sigma_3}\exp\left[-\frac{(y_i - \mu_3)^2}{2\sigma_3^2}\right]\right\}\right)$$
(8.22)

subject to
$$p_1 \geq 0,$$
$$p_2 \geq 0, \qquad (8.23)$$
$$(1 - p_1 - p_2) \geq 0.$$

The data for this are given in Table 8.3.

Use of SUMT algorithm with an initial starting point of

$$(p_1^\circ, p_2^\circ, \mu_1^\circ, \mu_2^\circ, \mu_3^\circ, \sigma_1^\circ, \sigma_2^\circ, \sigma_3^\circ) = (.1, .2, 100., 125., 175., 11.2, 13.2, 15.8)$$

yields estimates
$$\bar{p}_1 = .365$$
$$\bar{p}_2 = .475$$
$$\bar{p}_3 = .160 = (1 - \bar{p}_1 - \bar{p}_2)$$
$$\bar{\mu}_1 = 130.1$$
$$\bar{\mu}_2 = 163.1$$
$$\bar{\mu}_3 = 221.2$$
$$\bar{\sigma}_1 = 12.0$$
$$\bar{\sigma}_2 = 18.5$$
$$\bar{\sigma}_3 = 18.5.$$

Table 8.3 Data Giving Systolic Blood Pressure Values with Frequency of Occurrence

Systolic Blood Pressure	Frequency of Occurrence	Systolic Blood Pressure	Frequency of Occurrence	Systolic Blood Pressure	Frequency of Occurrence
95	1	150	17	200	3
105	1	155	4	205	3
110	4	160	20	210	8
115	4	165	8	215	1
120	15	170	17	220	6
125	15	175	8	225	0
130	15	180	6	230	5
135	13	185	6	235	1
140	21	190	7	240	7
145	12	195	4	245	1
				260	2

8.6 SOURCE OF PROBLEM AND REFERENCES

Source

We worked on this problem with Dale M. Heien of The George Washington University Logistics Research Project. The maximum likelihood estimation model and its data were given to us by E. A. Murphy of The Johns Hopkins University.

References

[1] V. G. Ashar and T. D. Wallace, "A Sampling Study of Minimum Absolute Deviations Estimators," *J. Operations Res. Soc. Am.*, **11**, 747–758 (1963).
[2] A. Charnes, W. W. Cooper, and R. O. Ferguson, "Optimal Estimation of Executive Compensation by Linear Programming," *Management Sci.*, **1**, 138–151 (1955).
[3] H. Cramér, *The Elements of Probability Theory*, Wiley, New York, 1955.
[4] A. J. Duncan, *Quality Control and Industrial Statistics*, Irwin, Homewood, Ill., 1955.
[5] W. D. Fisher, "A Note on Curve Fitting with Minimum Deviations by Linear Programming," *J. Am. Statist. Assoc.*, **56**, 359–362 (1961).
[6] A. S. Goldberger, *Econometric Theory*, Wiley, New York, 1964.
[7] A. A. Goldstein, N. Levine, and J. B. Hereshoff, "On the 'Best' and 'Least' Qth Approximation of an Overdetermined System of Linear Equations," *J. Assoc. Computing Machinery*, **4**, 341–347 (1957).
[8] H. O. Hartley, "The Modified Gauss-Newton Method for the Fitting of Nonlinear Regression Functions by Least Squares," *Technometrics*, **3**, 269–280 (1961).
[9] J. E. Kelley, Jr., "An Application of Linear Programming to Curve Fitting," *J. Soc. Ind. Appl. Math.*, **6**, 15–22 (1958).
[10] J. R. Meyer and R. B. Glauber, *Investment Decisions, Economic Forecasting, and Public Policy*, Division of Research, Harvard Business School, Boston, Mass., 1964.
[11] H. Theil, "On the Use of Incomplete Prior Information in Regression Analysis," *J. Am. Statist. Assoc.*, **58**, 401–415 (1963).
[12] H. M. Wagner, "Linear Programming Techniques for Regression Analysis," *J. Am. Statist. Assoc.*, **54**, 206–212 (1959).
[13] H. M. Wagner, "Non-Linear Regression with Minimal Assumptions," *J. Am. Statist. Assoc.*, **57**, 572–578 (1962).

9
DETERMINISTIC NONLINEAR PROGRAMMING EQUIVALENTS FOR STOCHASTIC LINEAR PROGRAMMING PROBLEMS

In this chapter we discuss the formulation and solution of a deterministic nonlinear programming problem that is equivalent to a stochastic linear programming problem of the chance-constrained type. In the chance-constrained problem the constraint coefficients are normally distributed random variables. The elements of the right-hand side and of the criterion function are deterministic. As an example of the application of this formulation, we discuss a chance-constrained programming model of minimum cost cattle feed under probabilistic protein constraints.

The chance-constrained programming problem is treated by Charnes and Cooper [1, 2]. They proposed [2] the deterministic equivalent presented in this chapter. Van de Panne and Popp [4] formulated the example problem and solved the associated nonlinear programming problem. Fiacco and McCormick [3] also solved it by using the sequential unconstrained minimization technique.

9.1 CHANCE-CONSTRAINED PROGRAMMING PROBLEM AND ITS DETERMINISTIC EQUIVALENT

The ordinary deterministic linear programming problem corresponding to the chance-constrained problem to be discussed herein can be written as follows. Determine x_j ($j = 1, \ldots, n$) to

$$\text{minimize} \sum_{j=1}^{n} c_j x_j \tag{9.1}$$

Chance-Constrained Problem and Deterministic Equivalent

subject to

$$\sum_{j=1}^{n} a_{ij} x_j \geq b_i, \quad i = 1, \ldots, m \tag{9.2}$$

and

$$x_j \geq 0, \quad j = 1, \ldots, n, \tag{9.3}$$

where a_{ij} ($i = 1, \ldots, m$ and $j = 1, \ldots, n$) are the constraint coefficients, the b_i's are elements of the right-hand side, the c_j's are elements of the criterion function, and the non-negative x_j's are the variables to be determined.

The chance-constrained formulation of the linear programming problem to be considered extends the deterministic problem given above as follows. Determine x_j ($j = 1, \ldots, n$) to minimize (9.3) subject to

$$P\left[\sum_{j=1}^{n} \tilde{a}_{ij} x_j \geq b_i\right] \geq \alpha_i, \quad i = 1, \ldots, m \tag{9.4}$$

and to (9.2). Here some or all of the coefficients $\tilde{a}_{ij}$ ($i = 1, \ldots, m$ and $j = 1, \ldots, n$) are random variables with normal distributions, and α_i ($i = 1, \ldots, m$) are prescribed probabilities with which the m constraints must be satisfied.

Independent Normal Case

Let us assume for the ith constraint that the $\tilde{a}_{ij}$'s are independent normal random variables with means

$$\bar{a}_{i1}, \ldots, \bar{a}_{in} \tag{9.5a}$$

and variances

$$\sigma^2(\tilde{a}_{i1}), \ldots, \sigma^2(\tilde{a}_{in}). \tag{9.5b}$$

Define for the ith constraint

$$\tilde{u}_i \equiv \sum_{j=1}^{n} \tilde{a}_{ij} x_j. \tag{9.6}$$

This variable is normally distributed with mean

$$\bar{u}_i = \sum_{j=1}^{n} \bar{a}_{ij} x_j \tag{9.7a}$$

and variance

$$\sigma^2(\tilde{u}_i) = \sum_{j=1}^{n} \sigma^2(\tilde{a}_{ij}) x_j^2. \tag{9.7b}$$

The ith constraint of the chance-constrained problem may be restated as

$$P[\tilde{u}_i \geq b_i] \geq \alpha_i. \tag{9.8}$$

We investigate the left side of inequality (9.8) to represent it in terms of x_j ($j = 1, \ldots, n$) and standard functions.

We write

$$P[\tilde{u}_i \geq b_i] = \int_{b_i}^{\infty} h(u_i)\, du_i, \tag{9.9}$$

where $h(u_i)$ is the normal density function of $\tilde{u}_i$. Setting

$$z_i = \frac{u_i - \bar{u}_i}{[\sigma^2(\tilde{u}_i)]^{1/2}} = \frac{u_i - \sum_{j=1}^{n} \bar{a}_{ij} x_j}{\left[\sum_{j=1}^{n} \sigma^2(\tilde{a}_{ij}) x_j^2\right]^{1/2}} \tag{9.10}$$

and substituting the lower limit of integration, we obtain

$$P[\tilde{u}_i \geq b_i] = \int_{\frac{b_i - \sum_{j=1}^{n} \bar{a}_{ij} x_j}{\left[\sum_{j=1}^{n} \sigma^2(\tilde{a}_{ij}) x_j^2\right]^{1/2}}}^{\infty} f(z_i)\, dz_i, \tag{9.11}$$

where $f(y)$ is the standardized normal density function

$$f(y) = \frac{1}{\sqrt{2\pi}} e^{-\frac{1}{2}y^2}. \tag{9.12}$$

In terms of the standardized normal left-tail cumulative function

$$P[\tilde{u}_i \geq b_i] = 1 - F\left\{\frac{b_i - \sum_{j=1}^{n} \bar{a}_{ij} x_j}{\left[\sum_{j=1}^{n} \sigma^2(\tilde{a}_{ij}) x_j^2\right]^{1/2}}\right\}. \tag{9.13}$$

We need to have $P[\tilde{u}_i \geq b_i] \geq \alpha_i$, so equivalently we need

$$F\left\{\frac{b_i - \sum_{j=1}^{n} \bar{a}_{ij} x_j}{\left[\sum_{j=1}^{n} \sigma^2(\tilde{a}_{ij}) x_j^2\right]^{1/2}}\right\} \leq 1 - \alpha_i, \tag{9.14}$$

and using the inverse function (also known as the percentage point or fractile) we obtain

$$\frac{b_i - \sum_{j=1}^{n} \bar{a}_{ij} x_j}{\left[\sum_{j=1}^{n} \sigma^2(\tilde{a}_{ij}) x_j^2\right]^{1/2}} \leq F^{-1}(1 - \alpha_i) \equiv \Psi(\alpha_i), \tag{9.15}$$

where $\Psi(\alpha_i)$ is the percentage point corresponding to $(1 - \alpha_i)$. If $\alpha_i = .95$, $\Psi(\alpha_i)$ is the .05 fractile. Finally, we can write the nonlinear constraint

$$\sum_{j=1}^{n} \bar{a}_{ij}x_j + \Psi(\alpha_i)\left[\sum_{j=1}^{n} \sigma^2(\tilde{a}_{ij})x_j^2\right]^{1/2} \geq b_i. \tag{9.16}$$

It is necessary to expand each of the stochastic constraints separately, considering the variances and required probabilities of satisfying the constraints α_i $(i = 1, \ldots, m)$.

Dependent Normal Case

Now let us consider the problem where for the ith constraint the $\tilde{a}_{ij}$'s are dependent multivariate normal. We use matrices and vectors because of the necessity to handle off-diagonal elements of the variance-covariance matrix. We define

$$\tilde{a}_i \equiv (\tilde{a}_{i1}, \ldots, \tilde{a}_{in})^t, \tag{9.17}$$

the distribution of which has mean

$$\bar{a}_i = (\bar{a}_{i1}, \ldots, \bar{a}_{in})^t \tag{9.18a}$$

and variance

$$V(\tilde{a}_i) = \begin{bmatrix} \text{var}(\tilde{a}_{i1}) & \cdots & \text{cov}(\tilde{a}_{i1}\tilde{a}_{in}) \\ \vdots & & \vdots \\ \text{cov}(\tilde{a}_{in}\tilde{a}_{i1}) & \cdots & \text{var}(\tilde{a}_{in}) \end{bmatrix}. \tag{9.18b}$$

Define for the ith constraint

$$\tilde{u}_i \equiv \tilde{a}_i^t x, \tag{9.19}$$

where

$$x = (x_1, \ldots, x_n)^t. \tag{9.20}$$

The random variable $\tilde{u}_i$ is normally distributed with mean

$$\bar{u}_i = \bar{a}_i^t x \tag{9.21a}$$

and variance

$$V(\tilde{u}_i) = x^t V(\tilde{a}_i)x. \tag{9.21b}$$

The ith constraint of the chance-constrained programming problem is again written

$$P[\tilde{u}_i \geq b_i] \geq \alpha_i.$$

Using the same procedure as previously, and substituting $\bar{a}_i^t x$ for $\sum_{j=1}^{n} \bar{a}_{ij}x_j$ and $x^t V(\tilde{a}_i)x$ for $\sum_{j=1}^{n} \sigma^2(\tilde{a}_{ij})x_j^2$, we obtain the constraint

$$\bar{a}_i^t x + \Psi(\alpha_i)[x^t V(\tilde{a}_i)x]^{1/2} \geq b_i. \tag{9.22}$$

9.2 EXAMPLE OF DETERMINISTIC EQUIVALENT FOR A CHANCE-CONSTRAINED PROGRAMMING PROBLEM

Determination of optimal computation of cattle feed is a well-known application of linear programming. The problem concerns the mixing of a number of raw materials in such a way that cattle feed is obtained that satisfies certain specified nutritive and other requirements with minimum cost for the input quantities of the raw materials. If the nutritive contents and unit costs of raw materials, and the requirements for nutrients, are known, the problem can be solved in a straightforward manner by linear programming methods. One problem that arises is that the nutritive content of the raw materials varies randomly, so that the solution given by linear programming using expected values, for instance, does not always satisfy the requirements. The example will deal with this type of problem. It is taken from van de Panne and Popp [4].

We first formulate the deterministic model assuming no random elements, and then go on to assume independent normal random variation in the protein content of the raw materials.

Deterministic Model

Table 9.1 gives the data of the problem. The percentage protein content and percentage fat content of the raw materials (barley, oats, sesame flakes, and groundnut meal) are given, with the required percentage content of protein and fat. Cost per ton of the four raw materials is also given. The problem is to determine a mix with minimum cost per ton that satisfies the nutritive requirements.

Let a_{ij} ($i = 1, 2$ and $j = 1, \ldots, 4$) denote the percentage protein content ($i = 1$) and the percentage fat content ($i = 2$) in the four raw materials. Let b_i ($i = 1, 2$) denote the percentage requirements. Let c_j ($j = 1, \ldots, 4$) denote the cost per ton of the raw materials. Let x_j ($j = 1, \ldots, 4$) note the fraction of the mixture that is composed of each of the raw materials. The deterministic linear programming model is as follows. Choose x_j ($j = 1, \ldots, 4$)

Table 9.1 Data for Deterministic Problem

	Barley	Oats	Sesame Flakes	Groundnut Meal	Requirement
Protein content (per cent)	12.0	11.9	41.8	52.1	21
Fat content (per cent)	2.3	5.6	11.1	1.3	5
Cost per ton (guilders)	24.55	26.75	39.00	40.50	

Example of Deterministic Equivalent

to
$$\text{minimize} \sum_{j=1}^{4} c_j x_j \tag{9.23}$$
subject to
$$\sum_{j=1}^{4} a_{ij} x_j \geq b_i, \quad i = 1, 2, \tag{9.24a}$$
$$\sum_{j=1}^{4} x_j = 1, \tag{9.24b}$$
$$x_j \geq 0, \quad j = 1, \ldots, 4. \tag{9.24c}$$

Writing out the constraints in detail using the data of the table, we obtain the following. Choose x_1, x_2, x_3, x_4 to

$$\text{minimize } 24.55 x_1 + 26.75 x_2 + 39.00 x_3 + 40.50 x_4 \tag{9.25}$$
subject to
$$12.0 x_1 + 11.9 x_2 + 41.8 x_3 + 52.1 x_4 \geq 21,$$
$$2.3 x_1 + 5.6 x_2 + 11.1 x_3 + 1.3 x_4 \geq 5,$$
$$x_1 + x_2 + x_3 + x_4 = 1,$$
$$x_1, x_2, x_3, x_4 \geq 0. \tag{9.26}$$

The optimal mixture solution of van de Panne and Popp [4] is as follows:

$$x_1 = .6852 \text{ (fraction barley)}$$
$$x_2 = .0127 \text{ (fraction oats)}$$
$$x_3 = .3021 \text{ (fraction sesame flakes)}$$
$$x_4 = 0 \quad \text{(fraction groundnut meal)}$$

with a cost of 28.94 guilders per ton.

Chance-Constrained Model with Deterministic Equivalent

This model differs from the previous one in two respects. First, the protein content of the four raw materials used for one batch of the mixture is constant but subject to variation for different batches of the mixture. The distribution of protein control of each raw material is normal and independent of the other raw materials, with mean equal to the values given previously and variance given below. Thus

$$\bar{a}_{11} = 12.0, \bar{a}_{12} = 11.9, \bar{a}_{13} = 41.8, \bar{a}_{14} = 52.1,$$
$$\sigma^2(\tilde{a}_{11}) = .28, \sigma^2(\tilde{a}_{12}) = .19, \sigma^2(\tilde{a}_{13}) = 20.5, \sigma^2(\tilde{a}_{14}) = .62.$$

Second, the specification is made that the probability of achieving a protein content of 21 per cent is at least .95.

Applying the procedure given in the previous section for a deterministic equivalent in the independent normal case, we obtain the following model.

Choose x_1, x_2, x_3, x_4 to

$$\text{minimize } 24.55x_1 + 26.75x_2 + 39.00x_3 + 40.50x_4 \qquad (9.27)$$

subject to

$$12.0x_1 + 11.9x_2 + 41.8x_3 + 52.1x_4,$$
$$+ (-1.645)[.28x_1^2 + .19x_2^2 + 20.5x_3^2 + .62x_4^2]^{1/2} \geq 21,$$
$$2.3x_1 + 5.6x_2 + 11.1x_3 + 1.3x_4 \geq 5,$$
$$x_1 + x_2 + x_3 + x_4 = 1,$$
$$x_1, x_2, x_3, x_4 \geq 0. \qquad (9.28)$$

The optimal mixture solution of van de Panne and Popp [4] is as follows:

$$x_1 = .6359 \text{ (fraction barley)}$$
$$x_2 = 0 \quad \text{(fraction oats)}$$
$$x_3 = .3127 \text{ (fraction sesame flakes)}$$
$$x_4 = .0515 \text{ (fraction groundnut meal)}$$

with a cost of 29.89 guilders per ton.

The increase in protein content required in the deterministic equivalent given by (9.27) and (9.28) over the stochastic programming problem given by (9.25) and (9.26) is satisfied by increasing the fraction of sesame flakes despite their high cost and high variance, and by introducing groundnut meal in place of oats along with a reduction in barley.

Fiacco and McCormick [3] obtained the same solution to the problem as van de Panne and Popp [4], using a different algorithm and also using a modified formulation where x_4 is replaced by $[1 - (x_1 + x_2 + x_3)]$.

9.3 SOURCE OF PROBLEM AND REFERENCES

Source

Fiacco, McCormick, and Mylander solved the problem using SUMT and presented the results in [3].

References

[1] A. Charnes and W. W. Cooper, "Chance-Constrained Programming," *Management Sci.*, **6**, 73–79 (1959).
[2] A. Charnes and W. W. Cooper, "Chance Constraints and Normal Deviates," *J. Am. Statist. Assoc.*, **57**, 134–148 (1962).
[3] A. V. Fiacco and G. P. McCormick, "Computational Algorithm for the Sequential Unconstrained Minimization Technique for Nonlinear Programming," *Management Sci.*, **10**, 601–617 (1964).
[4] C. van de Panne and W. Popp, "Minimum-Cost Cattle Feed under Probabilistic Protein Constraints," *Management Sci.*, **9**, 405–430 (1963).

10
OPTIMAL SAMPLE SIZES IN STRATIFIED SAMPLING ON SEVERAL VARIATES

In this chapter we discuss the use of nonlinear programming to determine optimal sample size allocations in stratified sampling problems when several variates are being sampled. The use of linear and nonlinear programming in approaching this type of problem has been discussed by several writers, for instance [2, 4–6], but primary concern has been given to reducing the direct model discussed herein into more simple mathematical programming models. For problems with two strata and several variates a graphical solution has been proposed by Dalenius [2], and Yates [8] has given a general mathematical approach useful for problems with a small number of strata and variates. However, formulation as a nonlinear programming problem and direct solution by general nonlinear programming algorithms has received less attention. Fiacco, McCormick, and Mylander [3] have treated the direct problem.

We use the notation of Cochran [1], and consider as an example problem one used by him, with four strata and two variates.

It is important to note in considering this problem that the real power of the nonlinear programming methods is seen when the number of variates grows large. The procedures given for determining the optimal sample sizes are very cumbersome, whereas nonlinear programming procedures can handle problems with many strata and many variates.

Decreasing marginal costs of sampling may be handled by branch and bound methods such as that discussed in Chapter 3.

We do not discuss optimal sample size in Bayesian stratified sampling. A discussion of the use of nonlinear programming techniques is given in Soland [7].

10.1 STRATIFIED SAMPLING PROBLEMS WITH NONLINEAR PROGRAMMING MODELS

The index h denotes the stratum and j the variate, where $h = 1, \ldots, L$ and $j = 1, \ldots, K$. We define

$N = $ total units in the population,
$N_h = $ total units in the hth stratum,
$n_h = $ number of units in the sample in the hth stratum,
$W_h = \dfrac{N_h}{N} = $ stratum weight.

The estimate of the population mean of the jth variate is

$$E(\bar{y}_j) = \frac{\sum_{h=1}^{L} N_h \bar{y}_{jh}}{N} = \sum_{h=1}^{L} W_h \bar{y}_{jh}, \tag{10.1}$$

where $\bar{y}_{jh}$ is the hth stratum mean of the jth variate. The variance of the estimate $\bar{y}_j$ is

$$V(\bar{y}_j) = \sum_{h=1}^{L} \frac{W_h^2 S_{jh}^2}{n_h} - \sum_{h=1}^{L} \frac{W_h S_{jh}^2}{N_h}, \tag{10.2}$$

as shown by Cochran [1], Chapters 5 and 5A, where S_{jh}^2 is the known sampling variance for the jth variate in the hth stratum.

In stratified sampling the values of the sample sizes, n_h, in the respective strata must be chosen by the sampler. They may be selected to minimize the variance of the estimate for a specified sampling cost, or to minimize the sampling cost for a specified value of the variance of the estimate. When only one variate is being sampled ($j = 1$), and the sample cost is of the form

$$C = c_0 + \sum_{h=1}^{L} c_h n_h, \tag{10.3}$$

optimal sample sizes for the various strata may be obtained by standard methods for either the minimum sampling variance criterion or the minimum cost criterion. When several variates are being sampled ($j > 1$) the problem is more difficult.

The optimal sampling problem is to minimize sampling cost subject to the constraints that the variance of the estimate of the population mean must be equal to or less than a specified value for all of the K variates. Constraints may be written as follows:

$$\sum_{h=1}^{L} \frac{W_h^2 S_{jh}^2}{n_h} - \sum_{h=1}^{L} \frac{W_h S_{jh}^2}{N_h} \leq V_j, \quad j = 1, \ldots, K, \tag{10.4}$$

where V_j is the upper limit on the variance of the estimate of the mean of the jth variate. It should be observed at this point that in the above constraints everything except the stratum sample sizes n_h ($h = 1, \ldots, L$) is known.

The sample size in the hth stratum must be non-negative, and it must be equal to or less than the total number of units in the stratum. Thus lower and upper bounds may be specified:

$$0 \leq n_h \leq N_h, \quad h = 1, \ldots, L. \tag{10.5}$$

Most of the references emphasize minimizing the linear sampling cost function, given by (10.3) above. The nonlinear programming model for obtaining optimal sample sizes with respect to this cost function is as follows.

Choose n_h ($h = 1, \ldots, L$) to minimize the linear cost function (10.3) subject to the nonlinear constraints (10.4) and to the non-negativity restrictions and upper bounds (10.5).

We discuss an example problem of this type in the next section. Kokan [5] has discussed this type of problem in some detail, and Jagannathan [4] has shown how it can be converted into one with linear constraints and nonlinear criterion function.

Certain types of sampling may have costs that are not linearly related to the number of units in the sample in the various strata. A more general cost function might be

$$C = c_0' + \sum_{h=1}^{L} c_h' n_h^p. \tag{10.6}$$

When the major element of cost is that of taking measurements on the unit, the previous cost function (10.3), where $p = 1$, may be appropriate. But when the major cost of sampling is a cost such as traveling between units, the relationship (10.6) with $p < 1$ (for instance, $p = .5$) may be more realistic, where c_h' is the travel cost between each unit. For the more general cost function, the following nonlinear programming problem would arise.

Choose n_h ($h = 1, \ldots, L$) to minimize the nonlinear criterion function (10.6) subject to (10.4) and (10.5). We do not give an example of this problem.

10.2 EXAMPLE WITH FOUR STRATA AND TWO VARIATES

The example is taken from Cochran [1, pp. 123–125]. The problem is to find the sampling plan with minimum cost where the variances of estimates of population mean for two variates are equal to or less than specified values. Table 10.1 gives data for the problem, including the population sizes in the strata, known variances of the two variates in the four strata, and unit cost in sampling in the four strata, where total cost equals $1 + \sum_{h=1}^{4} n_h$ (here $c_0 = 1$, $c_1 = 1, \ldots, c_4 = 1$).

Table 10.1 Data for Stratified Sampling Problem

Stratum h	Stratum Population N_h	Variances of jth Variate S_{j1}^2	S_{j2}^2	Cost per Unit of Sample c_h
1	400,000	25	1	1
2	300,000	25	4	1
3	200,000	25	16	1
4	100,000	25	64	1

The upper limits on the variances of estimates of the population means of the two variates are

$$V_1 \leq .04, \quad V_2 \leq .01.$$

The nonlinear programming problem is as follows. Choose n_1, n_2, n_3, and n_4 to minimize the linear criterion function

$$(1) + (1)(n_1) + (1)(n_2) + (1)(n_3) + (1)(n_4)$$

subject to

$$\frac{(.4^2)(25^2)}{n_1} + \frac{(.3^2)(25^2)}{n_2} + \frac{(.2^2)(25^2)}{n_3} + \frac{(.1^2)(25^2)}{n_4}$$

$$- \left[\frac{(.4)(25^2)}{400,000} + \frac{(.3)(25^2)}{300,000} + \frac{(.2)(25^2)}{200,000} + \frac{(.1)(25^2)}{100,000} \right] \leq .04,$$

$$\frac{(.4^2)(1^2)}{n_1} + \frac{(.3^2)(4^2)}{n_2} + \frac{(.2^2)(16^2)}{n_3} + \frac{(.1^2)(64^2)}{n_4}$$

$$- \left[\frac{(.4)(1^2)}{400,000} + \frac{(.3)(4^2)}{300,000} + \frac{(.2)(16^2)}{200,000} + \frac{(.1)(64^2)}{100,000} \right] \leq .01,$$

$$0 \leq n_1 \leq 400,000,$$
$$0 \leq n_2 \leq 300,000,$$
$$0 \leq n_3 \leq 200,000,$$
$$0 \leq n_4 \leq 100,000.$$

Solving the nonlinear programming problem given above, we obtain the optimal sample sizes and costs given in Table 10.2. The total cost is $1 + \sum_{h=1}^{4} c_h n_h = 232.2$. Cochran obtained the same sample sizes and costs using Yates' method.

Table 10.2 Optimal Sample Sizes and Stratum Sampling Costs

Stratum h	Optimal Sample Size n_h	Cost $c_h n_h$
1	193	193
2	180	180
3	187	187
4	171	181

10.3 SOURCE OF PROBLEM AND REFERENCES

Source

Fiacco, McCormick, and Mylander formulated and solved by the sequential unconstrained minimization technique the example problem of Cochran and presented the results in [3].

References

[1] W. G. Cochran, *Sampling Techniques*, 2nd Ed., Wiley, New York, 1963.
[2] T. Dalenius, *Sampling in Sweden: Contributions to the Methods and Theories of Sample Survey and Practice*, Almqvist and Wiksell, Stockholm, 1957.
[3] A. V. Fiacco, G. P. McCormick, and W. C. Mylander, III, "Nonlinear Programming, Duality and Shadow Pricing by Sequential Unconstrained Optimization," paper presented to First World Econometric Congress, Rome, September 14, 1965.
[4] R. Jagannathan, "The Programming Approach in Multiple Character Studies," *Econometrica*, **33**, 236–237 (1965).
[5] A. R. Kokan, "Optimum Allocation in Multivariate Surveys," *J. Roy. Statist. Soc.*, Ser. A, **126**, 557–565 (1963).
[6] S. Nordbotten, "Allocation in Stratified Sampling by Means of Linear Programming," *Skand. Aktuarietidskr.*, **39**, 1–6 (1956).
[7] R. M. Soland, "Optimal Multivariate Stratified Sampling with Prior Information," *Skand. Aktuarietidskr.*, **50**, 31–39 (1968).
[8] F. Yates, *Sampling Methods for Censuses and Surveys*, 3rd Ed., Griffin, London, 1960.

INDEX

Absolute deviations, 84
Acid dilution factor, 39
Advance planning, 60
Aerodynamic drag, 73
Airframe weight, 61
Alkylation problem, discussion of regression constraints, 5
 nonconcave constraints, 15
 nonseparability of, 16
Alkylation process optimization, 37
Ashar, V. G., 84
Atomic weights, 48

Bellman, R., 4
Bell-shaped exhaust chamber, 61
Bid evaluation, discrete characteristics, 4
Bid evaluation problem, description of, 28
 nonsmooth function in, 5
 separable objective function of, 15
Biomedical area, 91
Branch and bound techniques, discussion of, 30
 use of to handle nonsmooth functions, 5
Brush, M., 27
Burn time, 62
Burwick, C. W., 37
Butylene, 38

Carroll, C. W., 16
Cattle feed problem, constraints of, 15
 description of, 98
 nonseparability of, 16
 probabilistic aspects of, 5

CERs, 58, 79
Chance-constrained model with deterministic equivalent, 99
Chance-constrained programming, 5, 94
Chance-constrained programming problem, 97
Charnes, A., 84, 94
Chebyshev criterion, 85
Chemical equilibrium problem, convexity of, 15
 description of, 46
 nonseparability of, 16
Chemical process, 37
Cheney, E. W., 13
Cochran, W. B., 101
Colville, A. R., 37
Combinatorial problems, 4
 in bid evaluation, 30
Computer programs, 3
Concave function, 11
Constrained optimization problem, 1
Constraints, equality, 1
 inequality, 1
 nonlinear, 16
Continuous functions, 4
Controllable or knob variables, 38
Convex function, 11
Convex polyhedron, 7
Convex programming problem, analysis of models, 15
 description of, 11
 solution of by SUMT, 18
Convex set, 11

Cooper, W. W., 84, 94
Cost-estimating relationships (CERs), 58
Cost function, 59
Costing problems, 58
Cost models, 58
Cost prediction, 81
Cotton yarn characteristics, 86
Cramér, H., 91
Criterion function, 1
Cross-sectional area, 73
Curve fitting problem, classification of, 3
  description of, 83
  examples of, 2
  nonconvexity of, 15
  probabilistic aspects of, 5
Cutting plane method, 4, 13

Dalenius, T., 101
Dantzig, G. B., 3, 46
Davidon, W. C., 50, 57
Dependent variables, 38, 83
Derivatives, 30
Design parameters, 59
Design variables, 52, 60
Determination of weights, 69
Deterministic nonlinear programming
  equivalents, 94
Deterministic problems, 4, 98
Diminishing returns, 89
Discontinuous function, 30
Discrete problems, 4
Dual method, 33
Duncan, A. J., 86
Dynamic problems, 4

Earth-escape mission, 59
Eckhart, W., 27
Economically significant variables, 38
Engineering design problems, examples of, 2
  in bulkhead design, 50
  in launch vehicle design, 78
Equality constrained problems, 2
Equilibrium composition, 47
Extrapolation procedures, 18

Feasible directions, methods of, 14
Feasible region, 7
Feasible solution, 42
Ferguson, R. O., 84
F-4 performance number, 39

Fiacco, A. V., 14, 16, 94, 101
Fighter bombers, 24
Fisher, W. D., 84
Fixed-charge problem, 29
Flange, 52
Fletcher, R., 50, 57
Fractional damage, 23
Fractionator, 38
Free energy, 46
Fresh acid, 37

Game theory, 3
Gauss-Newton method, 89
Geometrical limitations, 55
Gibbs free energy function, 47
Glauber, R. B., 84
Global minimum, 10, 78
Goldberger, A. S., 84
Goldstein, A. A., 13, 86
Gradient projection method, 14, 16
Gramann, R., 27
Gravity-free vacuum, 75

Hadley, G., 10, 14
Hartley, H. O., 89
Heien, Dale M., 93
Hereshoff, J. B., 86
Hydrocarbon product, 37

Ideal velocity, 73
Impulse propellant weight, 61
Independent normal case, 95
Independent variables, 38, 39, 83
Individual engine thrust, 61
Inert weight, 61
Instrument unit weight, 62
Integer problems, 4
Intercontinental ballistic missiles, 24
Internal compartments of tankers, 50
Isobutane, 38
Isobutane makeup, 37
Isocontours, 7, 9

Jagannathan, R., 103
Johnson, S., 46
Jones, A. P., 36

Kavlie, D., 50, 57
Kelley, J. E., Jr., 13, 84
Kokan, A. R., 103

Index

Kowalik, J., 50, 57
Kuhn, H. W., 11

Lagrange multipliers, 2, 79
Large-scale decomposable systems, 5
Lateral walls, 50
Launch operations, cost, 61
Launch vehicle design, 58, 81
Learning curve, 63
Least squares, 10, 83
Levine, N., 86
Likelihood function, 90
Linear programming computer codes, 6
Linear programming problem, and bid evaluation model, 30
  detailed description of, 6
  general statement of, 1
  history of, 3
  separability of, 15
Linear regression problem, 86
Linear sampling cost function, 103
Liquid cargo, 50
Liquid oxygen/liquid hydrogen, 61
Liquid oxygen/rocket projectile, 61
Local minimum, 10, 78
Lockheed Missiles and Space Company, 59
Logarithmic transformation, 24
Logical problems, 4
Longitudinal bulkheads, 50
Long-range bombers, 24

McCormick, G. P., 14, 16, 94, 101
Marginal costs, 3, 32
Markowitz, H., 10
Mass balance constraints, 46
Mass fraction, 61
Material balances, 38
Mathematical programming problem, 1
Maximum likelihood estimation, 15, 90
Medium-range ballistic missiles, 24
Meyer, J. R., 84
Minimum absolute deviation estimators, 84
Minimum cost cattle feed, 94
Mitcherlisch's law, 89
Modal standard, 47
Moe, J., 50, 57
Moment of inertia, 54
Monte Carlo methods, 84

Motor octane number, 39
Murphy, E. A., 93
Mylander, W. Charles, III, 27, 101

National Aeronautics and Space Administration, 58
Nonconvex functions, 18
Nonconvex programming problem, 18, 78
Nonlinear programming problem, 1
Nonlinear relationships without discontinuities, 37
Nonnegativity restrictions, 103
Nonsmooth functions, 4
Nutritive contents, 98

Objective function, 1
Oil tanker, 50
Olefin feed, 37
Operating costs, 41
Operating ranges, 37
Operations cost, 66
Optimal control, 2
Optimal sample sizes, 101
Optimum-sized vehicle, 79
Optimum staging, 80

Panel, 52
Parameter estimation, 83
Partitioning, 30
Payload, 60
Payne, R. E., 38
Penalty factor, 17
Performance indices, 38
Performance parameter, 69
Petroleum industry, 37
Piecewise smooth, 30
Plate thickness, 55
Popp, W., 94, 98
Population mean, 102
Portfolio selection problem, 10
Positive semidefinite matrix, 7, 10
Powell, M. J. D., 50, 57
Pressure in atmospheres, 47
Principle of optimality, 4
Probabilistic problems, 4
Probabilistic protein constraints, 94
Process flow diagram, 37
Process variables, 38
Production cost, 63
Profit function, 37, 41

Propulsion, 61
Propulsive energy, 75
Pugh, R. E., 49

Quadratic programming problem, 7, 10, 15

Reactor, 37
Reactor acid strength, 39
Reactor temperatures, 39
Reduced gradient method, 14
Regression, 38, 41, 59
  linear, 15, 83
  nonlinear, 39, 83, 88
Regression coefficients, constrained, 84
Resource allocation, 3
Robust estimator, 84
Rosen, J. B., 14
Rush, Benjamin C., 81

Sample sizes, 102
Sampling plan, 103
Sauer, R. N., 37
Scaled costs, 29
Scheduling, 81
Section modulus, 53
Sensitivity, 44, 79
Separable programming problem, 15
Sequential unconstrained minimization technique (SUMT), 14, 16, 60
Setup cost, 29
Shadow prices, 79
Simplex method, 3, 10
Smooth functions, 4
Soland, R. M., 36
Space vehicles, 58
Specific impulse, 71
Spent acid, 37
Stage length, 70
Stage mass fraction, 70
Stage thrust, 69
Starting point sets, 77
Static problems, 4
Steepest descent, 2
Stochastic programming problems, 94

Stong, R. E., 18
Stratified sampling problem, 4, 15, 101
Stratum mean, 102
Structural complexity factor, 78
Structural integrity, 70
Structural optimization, 50
SUMT, 14, 33, 50, 57, 81, 100; *see also* Sequential unconstrained minimization technique
Supporting variables, 38

Target damage value, 22
Thrust of a single engine, 69
Thrust-to-initial-weight ratio, 72
Traveling salesman problem, 30
Tucker, A. W., 11

Unconstrained optimization problem, 2, 17, 91
Upper bounds, 103

Van de Panne, C., 94, 98
Variance of estimates of population mean, 103
Velocity, 60
Vendors, 28
Vertex, 6
Vertical deviations, 83
Vertically corrugated transverse bulkhead, 50

Wagner, H., 84
Wallace, T. D., 84
Weapons assignment problem, convexity of, 15
  description of, 22
  integer requirements of, 4
Web, 52
Weighted absolute vertical deviations, 86
Weighted expected damage, 24
White, W. B., 46

Yates, F., 101

Zoutendijk, G., 14